Hydrology - Current Research and Future Directions

Edited by Murat Eyvaz,
Ahmed Albahnasawi
and Motasem Y. D. Alazaiza

Published in London, United Kingdom

Hydrology - Current Research and Future Directions
http://dx.doi.org/10.5772/intechopen.110937
Edited by Murat Eyvaz, Ahmed Albahnasawi and Motasem Y. D. Alazaiza

Contributors
Abdelbagi Y. F. Adam, Ahmed Abdelmotlab Ahmed, Ahmed Albahnasawi, Ayan Saha, Deogratius Maiga, Dibyendu Saha, Didy Wurjanto, Erjola Keci, Juhani Pylkkanen, Juliane Castro Carneiro, Khuthadzo Mudzanani, Kushal Roy, Lidiane dos Santos Lima, Md Nazir, Mohamed B. O. Osman, Mohammed Ahmed Al-Ghobiri, Motasem Y. D. Alazaiza, Mpfunzeni Raphulu, Murat Eyvaz, Nilna Amal, Nyiko Maurice Chauke, Osman Ali Nasr, Paulo Cesar Colonna Rosman, Saeed Q. Al-Khalidi Al-Maliki, Sivuyisiwe Mapukata, Terence Phadi

First published in London, United Kingdom, 2024 by IntechOpen
IntechOpen is the global imprint of INTECHOPEN LIMITED, registered in England and Wales, registration number: 11086078, 167-169 Great Portland Street, London, W1W 5PF, United Kingdom

British Library Cataloguing-in-Publication Data
A catalogue record for this book is available from the British Library

Additional hard and PDF copies can be obtained from orders@intechopen.com

Hydrology - Current Research and Future Directions
Edited by Murat Eyvaz, Ahmed Albahnasawi and Motasem Y. D. Alazaiza
p. cm.
Print ISBN 978-0-85466-056-8
Online ISBN 978-0-85466-055-1
eBook (PDF) ISBN 978-0-85466-057-5

For EU product safety concerns:
IN TECH d.o.o., Prolaz Marije Krucifikse Kozulić 3, 51000 Rijeka, Croatia,
info@intechopen.com or visit our website at intechopen.com.

Meet the editors

Dr. Murat Eyvaz is an associate professor in the Environmental Engineering Department, Gebze Technical University, Turkey. His research interests are applications in water and wastewater treatment facilities, electrochemical treatment processes, filtration systems at the lab and pilot scale, membrane processes (forward osmosis, reverse osmosis, membrane bioreactors), membrane manufacturing methods (polymeric membranes, nanofiber membranes, electrospinning), spectrophotometric analyses (UV, atomic absorption spectrophotometry), and chromatographic analyses (gas chromatography, high-pressure liquid chromatography). He has co-authored many journal articles and conference papers and has taken part in many national projects. He serves as an editor for 50 journals and a reviewer for 120 journals. He has four patents on wastewater treatment systems.

Dr. Ahmed Albahnasawi is a research fellow in the Environmental Engineering Department, Gebze Technical University, Turkey. His graduate work focused on investigating the treatability of sequential anoxic-aerobic batch reactors followed by a ceramic membrane for textile wastewater treatment. Based on his Ph.D. research, Dr. Albahnasawi published three journal articles and participated in three international conferences. His current research interests are the application and design of microbial fuel cells integrated with Fenton oxidation for industrial wastewater treatment/solid waste management and monitoring of organic micropollutants by both chromatographic and spectrophotometric analyses.

Dr. Motasem Y. D. Alazaiza is an assistant professor in the Civil and Environmental Engineering Department, A'Sharqiyah University, Oman. His graduate work focused on the photographic methods techniques in monitoring non-aqueous phase liquid migration in porous medium. Based on his Ph.D. research, Dr. Alazaiza published numerous journal articles and participated in numerous international conferences. His current research interests include the transport of multiple fluid phases (water, oil, gases) in porous media and the mass transfer between those phases, with a particular specialization in the behavior of gases in contaminated groundwater systems. In addition, his research is applied to understand the fate and transport of non-aqueous phase liquids (NAPLs) in soil and groundwater, develop and improve remediation technologies, and assess risks associated with unconventional resource development, all using a combination of laboratory experiments and numerical modeling.

Contents

Preface

Hydrology – Current Research and Future Directions provides an overview of the behavior of water, which is our source of life, both on Earth and underground. It also provides an understanding of the systems water forms, discusses sustainable management of water, and examines methods and technologies required to transfer water to future generations. The book consists of nine chapters.

Chapter 1 looks at the history of hydrology and provides information about the use and management of water by past and present societies. Chapter 2 includes a study on tropical peat, one of the important elements of hydrology. In this field study conducted in Indonesia's peatlands, researchers collected data from local people and examined the interactions of society's land use dynamics and hydrological processes. Chapter 3 presents a water management study. Phytoremediation, a natural treatment of polluted water resources using plants, is being investigated for its potential use in the sustainable ecological restoration of abandoned mine sites in India's Raniganj Coalfield. In Chapter 4, researchers develop a model on how port infrastructure and water resources can be protected against the possible negative effects of climate change with a study they conducted in Brazil's Paraíba do Sul River using predictive algorithms and geographical technologies. They provide an example of how their proposed model can be used in the decision-making process against the challenges it poses. Chapter 5 considers the damage caused by water scarcity to people in the arid regions of Sudan in the context of integrated water resources management and sustainable governance. Chapter 6 focuses on another arid area, the Jazan region of Saudi Arabia. An environmental monitoring system using geographic information systems and remote sensing technologies is being established in this region to improve water dynamics negatively affected by climate change pressures. Chapter 7 highlights that water quality can be protected with ecological management in which water quality standards and water quality assessment methods are indispensable. In this context, the relationships of hydrological cycle elements are revealed. Chapter 8 includes a study of turning wastewater into a valuable resource via molecular activation and clarification reactions integrated with an innovative water purification approach. Finally, Chapter 9 presents new methodologies to restore ecological balance and recover water resources after acid mine drainage.

When we look at the research and approaches in various sections of the book, we notice again that water touches every element of Earth's life. We need to protect the water that connects us to life and each other, from water molecules to the great oceans, and leave a clean and sufficient legacy to future generations. We express our heartfelt gratitude to the research authors who contributed their invaluable insights and expertise to this book. We would also like to thank the referees and our editorial team who worked hard to bring the book to its final form. Of course, we are also grateful to you, our valued readers, who showed interest in this study. We hope that this book will inspire all readers, not just those working in hydrology, to take steps to protect water in the future.

Murat Eyvaz
Associate Professor,
Department of Environmental Engineering,
Gebze Technical University,
Kocaeli, Turkey

Dr. Ahmed Albahnasawi
Department of Environmental Engineering,
Gebze Technical University,
Kocaeli, Turkey

Motasem Y.D. Alazaiza
Assistant Professor,
Department of Civil and Environmental Engineering,
A'Sharqiyah University,
Ibra, Oman

Chapter 1

Introductory Chapter: Hydrology – Tracing the Past, Understanding the Present, Shaping the Future

Ahmed Albahnasawi, Motasem Y.D. Alazaiza and Murat Eyvaz

1. Introduction

Hydrology, the study of Earth's water systems, stands at the confluence of ancient wisdom and innovative science. This introductory chapter serves as a gateway into the expansive realm of "Hydrology: Current Research and Future Directions." In these pages, we embark on a journey that transcends temporal boundaries, tracing the historical currents that have shaped our understanding of water and navigating the contemporary landscapes of hydrological research. The historical roots of hydrology run deep, intertwining with the dawn of human civilization. From the earliest agricultural societies harnessing the power of rivers to the Renaissance-era pioneers formalizing hydrological principles, we uncover the threads that weave through time. As we delve into this historical tapestry, we recognize that our ancestors' reverence for water laid the foundation for a scientific discipline that now grapples with the complexities of climate change, urbanization, and sustainability.

Contemporary hydrology stands as a testament to human ingenuity and technological prowess. Advanced tools, from satellite-based remote sensing to sophisticated modeling techniques, empower researchers to unravel the intricacies of Earth's water dynamics. The present state of hydrology reflects a synthesis of diverse knowledge domains, where ecological considerations intersect with the precision of numerical models, offering a holistic understanding of water as a dynamic and interconnected force. Yet, with progress comes a challenge. The chapters within this volume unravel the multifaceted issues confronting hydrology in the twenty-first century. From the far-reaching impacts of climate change to the intricate dance between urban development and water cycles, each challenge presents an opportunity for exploration, understanding, and innovative solutions. As we stand on the precipice of unprecedented environmental shifts, hydrology emerges as a crucial lens through which we perceive and respond to the evolving dynamics of our planet. "Hydrology: Current Research and Future Directions" is more than a compendium of scientific discourse; it is an invitation to explore the intersections of knowledge, to navigate the complexities of our water systems, and to envision a future where sustainable water management is not just a goal but a shared responsibility. In the following chapters, leading experts guide us through advanced modeling techniques, ecological nuances, and the societal implications of hydrological research. Together, we embark on a collective endeavor to illuminate the path forward, where hydrology becomes a beacon guiding us to a more resilient and harmonious coexistence with water. As we turn the pages of this volume, we invite readers to join us in this exploration of hydrology's currents—past, present, and future.

2. Historical perspective of hydrology

The historical roots of hydrology are deeply embedded in the narrative of human civilization. Ancient societies recognized the vital role of water in sustaining life and agriculture, leading to early attempts at developing irrigation systems. However, it was during the Renaissance that hydrology started to emerge as a formalized scientific discipline. The seventeenth-century scientific revolution witnessed pivotal contributions from figures like Darcy and Horton, who introduced fundamental concepts like Darcy's law and provided insights into infiltration processes [1, 2].

Hydrology's historical progression reflects humanity's evolving understanding of the intricate processes governing water on Earth. The transition from qualitative observations to quantitative analyses marked a significant shift, providing a structured framework for investigating water-related phenomena and laying the foundation for the sophisticated hydrological models in use today.

3. Current trends in hydrology research

The latter half of the twentieth century and the early twenty-first century have seen a change in thinking in hydrology, propelled by technological advancements and a deeper understanding of Earth's interconnected systems. Remote sensing technologies, geographic information systems (GIS), and advanced numerical modeling techniques have revolutionized how hydrologists collect and analyze data [3]. These tools empower researchers to explore hydrological extremes, study watershed dynamics, and assess the impacts of human-induced changes on water resources [4]. The integration of satellite-based observations with hydrological models has become instrumental in continuously monitoring and predicting changes in water availability and quality [5].

Beyond technological frontiers, contemporary hydrology research increasingly recognizes the integral link between water systems and ecological processes. The emergence of ecohydrology as a distinct sub-discipline underscores the interconnectedness of water and ecosystems. This trend reflects a change in basic assumptions from viewing water merely as a resource to understanding it as a fundamental part of broader ecological systems. Riparian zones, wetlands, and aquatic ecosystems are focal points of hydrological research that integrates ecological principles. Researchers investigate how vegetation influences water flow patterns, nutrient cycling, and the overall health of aquatic environments [6]. This holistic approach not only expands our understanding of hydrological processes but also underscores the importance of preserving ecosystem integrity for sustainable water management.

3.1 Human–water interactions

In an era marked by rapid urbanization and population growth, the intersection of hydrology with human activities has become increasingly significant. Urban hydrology investigates the complex dynamics of water in urban landscapes. The replacement of natural surfaces with impervious ones, coupled with the heat island effect, alters precipitation patterns, runoff characteristics, and overall hydrological processes [7]. Understanding these interactions is vital for mitigating urban flooding, optimizing stormwater management, and ensuring the resilience of urban water systems.

Intensified agricultural practices, driven by the need to feed a growing global population, also play a central role in shaping contemporary hydrology trends. Researchers explore sustainable agricultural practices, water use efficiency, and the impact of agricultural runoff on water quality [8]. This intersection of hydrology with agriculture emphasizes the need for holistic, adaptive approaches to address the challenges of feeding the world while preserving water resources.

3.2 Global perspectives and collaborative initiatives

The trends in hydrology research extend beyond regional boundaries, reflecting a global perspective on water-related challenges. Collaborative initiatives, often facilitated by international organizations, aim to address transboundary water issues, share data, and foster a collective understanding of global hydrological patterns [9]. These collaborative efforts recognize that water-related challenges, whether driven by climate change, land use changes, or demographic shifts, need cooperative solutions that transcend political borders.

4. Challenges in contemporary hydrology

As hydrology strides into the twenty-first century, it faces formidable challenges. Climate change, characterized by shifts in precipitation patterns and an increase in extreme weather events, poses a critical threat to water resource management8. The rapid pace of urbanization and intensified agricultural practices contribute to disruptions in hydrological cycles, resulting in issues such as urban flooding and contamination of water sources [10]. Achieving sustainable water resource management requires a comprehensive understanding of the complex interplay between human activities, climate variability, and ecosystem health. The depletion of groundwater aquifers, a cornerstone of water security, demands innovative solutions [11]. In this context, the integration of artificial intelligence (AI) and machine learning (ML) emerges as a promising avenue, offering tools to model complex hydrological processes with greater accuracy and predictability [12].

The challenges faced by contemporary hydrology are multifaceted and require interdisciplinary solutions. Climate change, driven by anthropogenic activities, has led to a heightened frequency and intensity of extreme weather events. These shifts impact precipitation patterns, snowmelt, and overall water availability [13]. The ramifications extend beyond mere fluctuations in water quantity, affecting water quality and ecosystem health.

Urbanization, often considered a hallmark of progress, poses a unique set of challenges to hydrologists. Rapid urban development alters natural landscapes, replacing permeable surfaces with impervious ones, leading to increased runoff and heightened flood risks [14]. The intricate balance of urban hydrology is further complicated by the heat island effect, where urban areas experience higher temperatures than their rural counterparts, influencing evaporation rates and water demand. Intensified agricultural practices, essential for feeding the burgeoning global population, have their own hydrological footprint. Excessive irrigation, fertilizer runoff, and land use changes contribute to soil erosion and water contamination, need sustainable agricultural practices [15]. The complex interplay of urbanization, agriculture, and climate change underscores the need for holistic, integrated approaches in contemporary hydrological research.

5. Future directions in hydrology research

Looking ahead, the future of hydrology promises exciting possibilities as the field continues to evolve in response to ongoing challenges and technological advancements. The integration of artificial intelligence (AI), machine learning (ML), and big data analytics stands out as a transformative force, poised to enhance predictive modeling and decision-making in water resource management. AI and ML algorithms, equipped to analyze extensive datasets, hold the potential to unravel intricate patterns within complex hydrological systems [16, 17].

Interdisciplinary collaboration is a key driver for future progress in hydrology, as the field intersects with ecology, climate science, and engineering. Holistic approaches, encompassing social, economic, and environmental considerations, have become imperative in shaping sustainable water management strategies. The challenges posed by climate change, population growth, and environmental degradation need innovative solutions that transcend traditional disciplinary boundaries, fostering collaboration among scientists, policymakers, and stakeholders [18].

6. Conclusion

In conclusion, "Hydrology – Current Research and Future Directions" aspires to be more than a compendium of scientific knowledge. It looks to inspire a collective effort toward understanding and addressing the profound challenges posed by Earth's water systems. By tracing the historical roots, elucidating current trends, and envisioning future trajectories, the book invites readers on a journey beyond the laboratory and into policy, society, and the environment.

As we stand on the precipice of unprecedented environmental change, the importance of hydrology in shaping our response cannot be overstated. The book, with its multifaceted exploration, aims to be a catalyst for a change in thinking in how we perceive, study, and manage water resources. In doing so, it lays the groundwork for a future where hydrology, informed by collaboration, innovation, and a deep understanding of Earth's interconnected systems, plays a pivotal role in building a sustainable and resilient world.

Author details

Ahmed Albahnasawi[1], Motasem Y.D. Alazaiza[2] and Murat Eyvaz[1*]

1 Department of Environmental Engineering, Gebze Technical University, Kocaeli, Turkey

2 Department of Civil and Environmental Engineering, A'Sharqiyah University, Ibra, Oman

*Address all correspondence to: meyvaz@gtu.edu.tr

References

[1] Darcy H. Les Fontaines Publiques de la Ville de Dijon. Paris: Dalmont; 1856

[2] Horton RE. The Rôle of infiltration in the hydrologic cycle. Eos, Transactions American Geophysical Union. 1933;**14**(1):446-460. DOI: 10.1029/TR014i001p00446

[3] Kushwaha N, Elbeltagi A, Patel A, Zakwan M, Rajput J, Sharma P. Assessment of water resources using remote sensing and GIS techniques. Current Directions in Water Scarcity Research. 2021;7:85-98. DOI: 10.1016/B978-0-323-91910-4.00006-6

[4] Moreno HA, Ogden FL, Alvarez LV. Unstructured-mesh terrain analysis and incident solar radiation for continuous hydrologic modeling in mountain watersheds. Water. 2018;**10**(4):398. DOI: 10.3390/w10040398

[5] Solakian J, Maggioni V, Lodhi A, Godrej A. Investigating the use of satellite-based precipitation products for monitoring water quality in the Occoquan watershed. Journal of Hydrology: Regional Studies. 2019;**26**:100630. DOI: 10.1016/j.ejrh.2019.100630

[6] Cao Z, Wang S, Luo P, Xie D, Zhu W. Watershed ecohydrological processes in a changing environment: Opportunities and challenges. Water. 2021;**14**(9):1502. DOI: 10.3390/w14091502

[7] Vujovic S, Haddad B, Karaky H, Sebaibi N, Boutouil M. Urban heat island: Causes, consequences, and mitigation measures with emphasis on reflective and permeable pavements. Civil Engineering. 2021;**2**(2):459-484. DOI: 10.3390/civileng2020026

[8] Yimer EA, Riakhi F, Bailey RT, Nossent J, Van Griensven A. The impact of extensive agricultural water drainage on the hydrology of the Kleine Nete watershed, Belgium. Science of the Total Environment. 2023;**885**:163903. DOI: 10.1016/j.scitotenv.2023.163903

[9] van Hateren TC, Jongen HJ, Al-Zawaidah H, Beemster JGW, Boekee J, Bogerd L, et al. Where should hydrology go? An early-career perspective on the next IAHS scientific decade: 2023-2032. Hydrological Sciences Journal. 2023;**68**(4):529-541. DOI: 10.1080/02626667.2023.2170754

[10] Fatichi S, Vivoni ER, Ogden FL, Ivanov VY, Mirus B, Gochis D, et al. An overview of current applications, challenges, and future trends in distributed process-based models in hydrology. Journal of Hydrology. 2016;**537**:45-60. DOI: 10.1016/j.jhydrol.2016.03.026

[11] National Research Council. Challenges and Opportunities in the Hydrologic Sciences. Washington, DC: The National Academies Press; 2012. DOI: 10.17226/13293

[12] Yao S, Chen C, Chen Q, Zhang J, He M. Combining process-based model and machine learning to predict hydrological regimes in floodplain wetlands under climate change. Journal of Hydrology. 2023;**626**:130193. DOI: 10.1016/j.jhydrol.2023.130193

[13] Muller M. Three challenges to keep hydrologists and planners busy for another decade. In: 19th SANCIAHS Symposium - South African National Committee of the International Association of Hydrological Sciences, 18-20th September 2018. Skukuza:

Nombolo Mdluli Conference Centre, Kruger National Park; 2018. Available from: https://ssrn.com/abstract=3287160

[14] Feng B, Zhang Y, Bourke R. Urbanization impacts on flood risks based on urban growth data and coupled flood models. Natural Hazards. 2021;**106**:613-627. DOI: 10.1007/s11069-020-04480-0

[15] Mateo-Sagasta J, Marjani Zadeh S, Turral H, Burke J. Water Pollution from Agriculture: A Global Review - Executive Summary. Rome, Italy: Food and Agriculture Organization of the United Nations (FAO); 2017. International Water Management Institute (IWMI), CGIAR Research Program on Water, Land and Ecosystems (WLE). Colombo, Sri Lanka, 35p

[16] Quilty J, Adamowski J, Boucher A. A stochastic data-driven ensemble forecasting framework for water resources: A case study using ensemble members derived from a database of deterministic wavelet-based models. Water Resources Research. 2018;**55**(1):175-202. DOI: 10.1029/2018WR023205

[17] Shen C. A transdisciplinary review of deep learning research and its relevance for water resources scientists. Water Resources Research. 2018;**54**(11):8558-8593. DOI: 10.1029/2018WR022643

[18] Rangecroft S, Rohse M, Banks EW, Day R, Di Baldassarre G, Frommen T, et al. Guiding principles for hydrologists conducting interdisciplinary research and fieldwork with participants. Hydrological Sciences Journal. 2021;**66**(2):214-225. DOI: 10.1080/02626667.2020.1852241

Chapter 2

Applying Field Research in Tropical Peatlands and Managing Ground Data: A Challenging Approach

Nilna Amal and Didy Wurjanto

Abstract

Hydrology research of peatlands requires accurate data and field research, which is still challenging, especially in Indonesia's tropical peatlands. Peatlands in Indonesia have been explored and studied, including the peat soil and the community next to the place, requiring wise management. This study covers the challenge of conducting field research in two types of different land characteristics and societies. Hence, it explores how to present the field data. The result shows an appropriate approach to the community by understanding their background is essential. Four years of research in different peatland areas has proven this in either Sumatra Island or Kalimantan Island. The field research data can be presented in short or long intervals. The short data can show crucial details about rainfall-water table elevation (WTE) on one side. On the other hand, the long interval data can present a whole trend rainfall-WTE relationship and the possibility of fire occurrence and indicate the peatland condition, whether it is improving or deteriorating. Different results from both lands show WTE risk, which is indicated by the WTE numbers more than 40 cm below the surface reach exceeding 70%; it can be caused by different land use and intensity of activities.

Keywords: appropriate approach, hydrology study, land use, parameters interconnectivity, tropical peatland, water table elevation (WTE)

1. Introduction

Peatlands in the world only cover 3% of land and surface, but they store one-third of the total world's soil carbon, making them essential to keep carbon from being released into the atmosphere. Peatlands serve several roles, such as carbon storage and sink, locations for various chemical processes, and water retention, and sometimes contain vegetation and wildlife that differ from other upland ecosystems [1]. Peatlands are located in various regions, both in sub-tropical areas such as in northern Europe and North America, and tropical areas, such as Asia, especially Indonesia. The subject of the studies covers many aspects, from the management of the peat soil, and identification of the damage of the peatland usage to the restoration after disturbance and identifying conditions susceptible to peatland fires possibility [2–5].

As the largest tropical peatlands area, Indonesia has been the subject of worldwide studies for years, particularly those related to the Mega Rice Project in Central

Kalimantan. Although it was recognized as an unsuccessful program, it opened many eyes. It is realized that peatland ecosystems are not only marginal due to their limited properties but also functional areas when approached with a suitable program. The inappropriate approach resulted in several issues, such as soil peat drying, which increases the possibility of fire occurrences and decreases or changes the composition of organisms within that ecosystem. Several parameters are believed to be indicators for identifying damage possibilities. Those are water table elevation (WTE), precipitation, evapotranspiration, and soil moisture [2, 6–8]. These parameters need to be required as physical properties, and the process, particularly in Indonesia, presents challenges. Unlike in developed countries, meteorological data collected and managed by government institutions are retained for their own purposes and only published as analyses, making it difficult to obtain crucial data like hourly precipitation records or water elevation records at channels. Furthermore, obtaining advanced data such as water table elevation had not been easy in the past, but recently, it has become possible to acquire such data for specific regions. This situation discourages researchers from finding solutions to the challenges it presents. Therefore, this chapter aims to investigate how to conduct field research on the various land-use characteristics of peatlands and how to present the data in a way that is accessible to the stakeholders.

2. Research methodology

Data on hydrology variables in peatlands is available in both primary and secondary sources. The preliminary data can only be obtained by applying specific equipment at the study location. In the case of tropical peatlands, there are two alternative approaches: a disturbed or undisturbed peatland. A disturbed peatland is easier to access since it often has established paths and is usually located near communities. The main concern is only how to transport the equipment to the location and ensure its safety, even in the presence of many people. The appropriate approach to the nearest community will determine the equipment's safety. Hiring someone who oversees the installation at times may be the best solution, but it also entails allocating a specific budget for this purpose. Data collection in Padang Island began in 2019. Field research in the Gambut District of South Kalimantan was conducted in 2022, while it was completed in Barito Kuala, South Kalimantan, in 2021. Below, we explain how the field research on Padang Island Riau Province and South Kalimantan was conducted.

Padang Island is an island in the Malacca Strait, a peatland island located near Sumatra Island. The entire island is predominantly composed of peat soil, surrounded by the ocean. Padang Island is one of a series of four main islands within the Meranti Archipelago Regency, Riau Province. Geographically, Meranti Archipelago Regency is located at 0°42′30″-1°28′0”N and 102°12′0″-103°10′0″E. The climate in Padang Island and the Meranti Archipelago Regency is temperate, with maximum air temperatures ranging from 25–32°C. The details can be seen in **Figure 1a**. The location of research in South Kalimantan can be seen in **Figure 1b**. The Meranti Archipelago is situated on the eastern coast of Sumatra Island, with a coastline that borders neighboring countries. Padang Island, one of the peat islands in Riau province, where almost the entire island consists of peat, has been inhabited by the community since the late 19th century. Padang Island's topography is generally flat, with a maximum elevation of 15 m above mean sea level (MSL), and nearly all settlements are situated at altitudes lower than 6 m above MSL. Several researchers have conducted their study here [9, 10].

As shown in **Figures 1a** and **2a,b**, reaching Padang Island requires effort. Access to Padang Island is only possible by boat, and the journey takes approximately 45 minutes from the edge of Sumatra Island to the shore of Padang Island using a local boat. The local boat that may be used is depicted in **Figure 2a** and **b**. **Figure 2a** and **b** illustrates the traditional methods of reaching the location. Nevertheless, several researchers have conducted studies there, as indicated by the following references [9, 11]. The Indonesian government, through Peatland Restoration Agency-Badan Restorasi Gambut (BRG), now known as Peatland and Mangrove Restoration Agency-Badan Restorasi Gambut dan Mangrove (BRGM) – of the Republic of Indonesia,

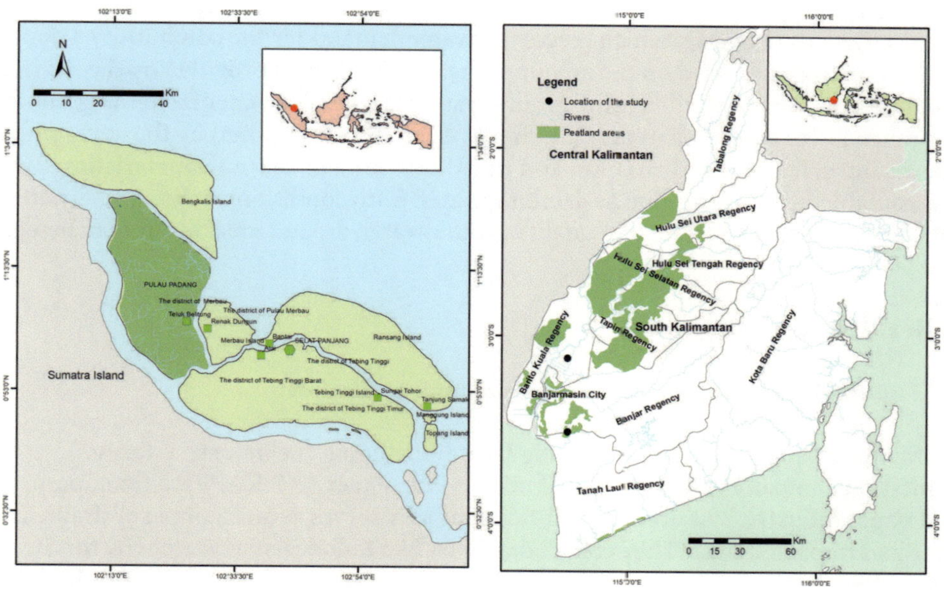

Figure 1.
(a) The location of Padang Island Riau Province and (b) the location of South Kalimantan, both in Barito Kuala and Banjar regency.

(a) (b)

Figure 2.
(a) The location of the local port on Sumatra edge and (b) the location of the local port on Padang Island edge.

has selected it as a pilot project for monitoring. However, there have been only a few changes there.

People around – just like most Indonesians – have rice as their primary food, but they do not have any farming land, so their needs will be provided by other regions outside of the island. The leading factor causing this is either the lack of knowledge of how to implement rice farming on peatland soil or the lack of experience in adopting a farmer's lifestyle. So, even though it would be better to try to obtain farming land, they will have enough land only for growing vegetables and crops instead, and they will import rice from outside the island.

The water for drinking and other domestic needs is provided by rainfall. They do not have a regional company like other Indonesian areas, namely Perusahaan Air Minum Daerah (PDAM), which serves the water demand for the community. People on Padang Island depend on the amount of water that can be provided by the environment; otherwise, they will use the groundwater storage. The water from the ground has much lower quality than the drinking water standard required by the government. For instance, it has a reddish color, and its pH is 4. Both of these characteristics make it unsuitable for consumption as drinking water. Both conditions, relying solely on rainfall for water and using low-quality groundwater, are not ideal for human beings.

3. Result and discussion

3.1 The challenge in applying the equipment installation

Many companies operate in Padang Island, including the industrial forest plantation company (HTI) PT Riau Andalan Pulp Paper (PT. RAPP), a fiber, pulp, and paper industry. The area around this company serves as an example of drained peatland impacted by it. They collect data, just like Indonesian researchers, for their respective purposes. They have installed their equipment and have been monitoring it for a while. Typically, they collect water table elevation data hourly and observe daily fluctuations in water elevation within the channels. The way they monitor the devices and evaluate the data can be seen in **Figure 3a** and **b**. At a particular time, the data can be obtained under specific conditions, exemplified by the collaboration in 2017 between the team from the Engineering Faculty of Universitas Gadjah Mada and BRG. Nevertheless, a similar opportunity did not arise in 2019 when we visited the location to try to obtain their data.

They refused to share their data since the researcher came without authority, even though the letter of recommendation was officially complete. They have been collecting such prestigious data that would not be shared with anyone unless it would provide an advantage for feedback. In 2017, there were no alternatives other than to hand it to the team. It was because the moratorium would be applied when the evidence showed that their activities were causing harm to the environment. The location and the route to access the location of their logger data can be seen in **Figure 3a**, and the location to measure the manual data for water elevations can be seen in **Figure 3b**. Their data logger is situated in the middle of the blocks which are filled with plantations such as *Acacia Crasicarpa*. The canal is distinct from those built in the community. The canal operated by the company is much larger, measuring about six to seven meters in width [12], whereas those in the community are only about two meters wide, as depicted in **Figure 4a** and **b**.

DOI: http://dx.doi.org/10.5772/intechopen.1004818

(a) (b)

Figure 3.
(a) The location of PT. RAPP data logger and (b) the location of manual water elevation observer.

(a) (b)

Figure 4.
(a) Canals in the company location and (b) the ditch in the community location.

Installing equipment in a peatland location necessitates the adoption of both a mechanical and a social approach. Researchers face various challenges when collecting primary data. One example is when the researchers came there, they were holding the local election at the same time. The event can affect either the process of taking data or the safety of the equipment. Therefore, adopting an approach that allows for effective engagement with the community and ensures unbiased interaction becomes paramount, even when the environmental conditions are generally unsupportive. However, when the equipment is installed adjacent to a company, its safety becomes contingent upon the company's policies regarding cooperation or refusal. This issue was exemplified by the loss of the last piece of equipment installed in Sei Hiu, located in close proximity to the HTI, shortly after commencing the monitoring of groundwater elevation or water table elevation (WTE) data.

Establishing connections with local communities also requires budget allocation. These hypotheses apply equally in the Gambut District, where the research was conducted there. The locals who support the research can be seen in **Figure 5a** and **b**. Furthermore, the area has attracted various farmer activities [13], indicating that it is not an entirely intact peatland. Gambut District is situated in South Kalimantan within Kalimantan Island. It is a part of Banjar Regency and is located adjacent to Banjarbaru City, which is poised to become the future capital of South Kalimantan. This strategic location makes it more accessible than Padang Island. However, the installation of automatic equipment poses budgetary challenges, making the manual option the preferred choice.

Applying water table equipment on a national scale requires regulations suitable for most regions, considering both the physical and the community conditions. BRGM has considered these needs and established the principles. This includes defining the regions with the highest priority for implementation and determining the appropriate number of equipment installations.

The implementation of peatland restoration is necessary to be able to be monitored in terms of its effectiveness in keeping the groundwater at the required levels. Groundwater level is the height of the groundwater level in peatland as measured from the peat surface to the peat groundwater level. One of the successes of peat restoration is if the damaged and dry peatland has become wet again, which means it has a water content reaching 0.4 meters from the surface. How do we monitor the groundwater level? The groundwater level can be monitored using a special infrastructure, namely, the groundwater level monitoring infrastructure (GWLMI). The

(a) (b)

Figure 5.
(a) Local community who supported the field research in Padang Island, (b) local community who supported the field research in Gambut District.

GWLMI is also an infrastructure or instrument installed to monitor the hydrological conditions of a Peatland Hydrological Unit (PHU).

The agreement is provided for each Hydrological Response Unit (HRU), which is different from PHU. The Peatland Hydrological Unit, referred to as Kesatuan Hidrologi Gambut in Indonesian language (KHG), is the Peat Ecosystem which is located between 2 (two) rivers, between the river and the sea and/or in a swamp. The Hydrological Response Unit (HRU) is the location where the GWLMI is built. The HRU is the smallest peat restoration management unit that has the same hydrological characteristics and is limited by peat ridges and/or canals that influence the water system in the sub-KHG or KHG area. While a river borders a KHG, HRUs are bordered by canals.

The GWLMI are situated and spread out at the targeted. The government of Indonesia, through the BRGM, has more than 150 infrastructures to monitor the groundwater level at priority locations throughout seven provinces in Indonesia. The GWLMI may consist of several components, including (a) sensor for measuring rod or water level, (b) soil moisture sensor, (c) rainfall sensor, (d) data loggers and/or (e) other supporting equipment. A team was assigned to build the GWLMI Installation: First comes first, the team carries out a field validation to ensure the suitability of the GWLMI installation location, by paying attention to, among other things, accessibility, security, mobile network, and local community approval.

The emphasis is likely determined based on several parameters that represent the peatland's condition and the equipment's need. These parameters measured include the density of channels, soil moisture, the extent of forest and land fires, and whether the peatland is swallow or deep peat. Different scoring will result in different priorities for the needed equipment, with the highest score defining the top priority for implementation. The one that receives the highest score will be defined as the highest priority for application. An example of the result of this method is shown in **Table 1**.

In **Table 1**, researchers can observe that nine pieces of equipment were installed around South Kalimantan, and this number may change based on the recent evaluation of new scoring for the region.

Building the GWLMIs to monitor the peatland groundwater level is part of the Government of Indonesia's efforts to restore the degraded peatlands. The activities of BRGM include building infrastructure for rewetting the degraded peatlands

Code	Regency	District	Village
BRG_630708_01	Hulu Sungai Tengah	Batang Alai Utara	Haur Gading
BRG_630805_01	Hulu Sungai Utara	Amuntai Tengah	Pinang Habang
BRG_630805_02	Hulu Sungai Utara	Amuntai Tengah	Mawar Sari
BRG_631104_01	Balangan	Batu Mandi	Mantimin
brg16	Hulu Sungai Utara	Banjang	Pulau Damar
BRG_630411_01	Barito Kuala	Kuripan	Jambu
BRG_630611_01	Hulu Sungai Selatan	Daha Barat	Baru
BRG_630804_01	Hulu Sungai Utara	Danau Panggang	Sungai Namang
BRG_630801_01	Hulu Sungai Utara	Amuntai Selatan	Murung Panggang

Table 1.
The gauges properties around South Kalimantan peatland areas.

such as canal-blockings, deep wells, and back-filling to stop canals from draining the peatland. Revegetation has also been conducted by BRGM to prevent peatland decomposition and reduce the emissions through land cover improvement. BRGM also provides economic incentives for local communities that contribute to protecting the surrounding peatland and utilize the wet peatland without draining it. BRGM also facilitates the operation of weather modification techniques to develop artificial rain during the dry season.

3.2 Water tale elevation data presentation

Water table elevation (WTE) has been defined as an essential indicator for a peatland. It determines the condition of the peatland and has become a focal point of research. Some researchers who conducted studies in this field covered degraded and undegraded peatlands [10, 14–17] and accepted it as a parameter that needs to be continuously monitored. As a variable, it is presented in various ways, either in interval monitoring or presenting methodology. Observation intervals can range from 10 minutes to 24 hours, so the way to present depends on the data provided. Researchers have utilized intervals of 10 minutes, 15 minutes, 30 minutes, and 1 hour [9, 10, 16, 18]. Different intervals have also been used, such as 3 hours [16], daily [2, 19], weekly and monthly [14, 15], and three times a month [19]. The data can be visualized in **Figure 6a** and **b**.

The interval data in **Figure 6a** and **b** show a relatively flat trend as it represents a condition on a particular day. However, slight variations can be observed within short time frames. For instance, in **Figure 6a**, the WTE started to increase at 03.20 and then remained relatively stable without significant fluctuations. Furthermore, the rain gauge data indicated that there was rainfall at 03.10. Additionally, these figures can be used to illustrate the rainfall-WTE relationships as well as the flow generated-WTE connection. The short-interval data can also reveal the generated flow and its influence on WTE [18, 20]. A similar pattern is evident in **Figure 6b** and **c**. It can be noticeable that data from different days exhibits similar trends. Consequently, it can be concluded that daily WTE tends to remain stable, except when influenced by rainfall events.

The number of data points within a specific interval can be presented as water table frequency, representing the percentage of data recorded. The water table frequency in Barito Kuala at Jambu Gauge compared to government regulations, which stipulate a maximum of 400 mm, can be seen in **Figure 7**.

Figure 7 shows that water table elevation may reach depths of up to 580 cm, which could pose a high risk. However, since it occurred in less than 2% of cases, it was scarce and may not be considered dangerous. The results showed that even though the maximum depth is very low, it generally complies with the government requirement of a maximum depth of 40 cm below the surface. Depths exceeding 40 cm occur in less than 5% of cases. When compared to the previous study on Padang Island [10], which was also conducted in tropical areas, the results indicated better conditions. In Padang Island, even in the Forest Zone area, the percentage of WTE exceeding the government regulations reached up to 80%. Several specific conditions may determine the condition of both.

Padang Island in Riau Province is an area where large companies operate their businesses and manage some industrial plantations that serve as a source of pulp paper. This situation significantly impacts water table elevation (WTE), causing it to become deeper in the vicinity of the company despite the area being

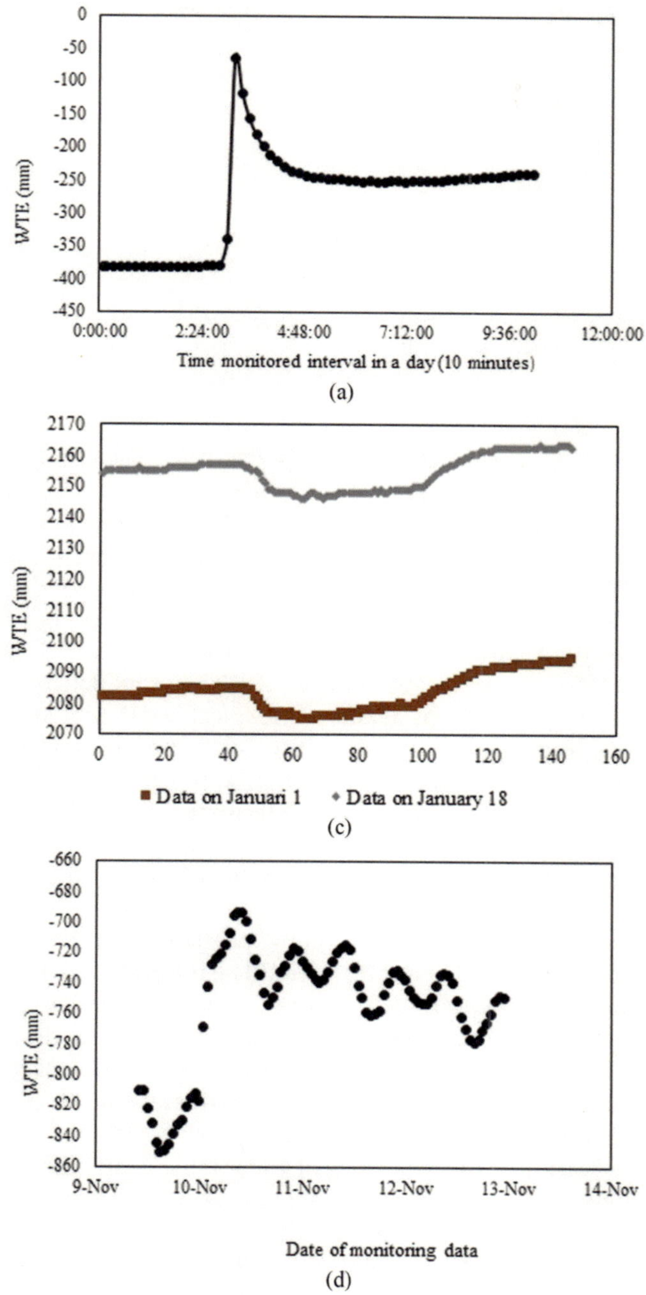

Figure 6.
(a) WTE fluctuation in a day with 10-minute interval monitoring. (b) WTE fluctuation in a day with 10-minute interval monitoring at Padang Island. (c) WTE fluctuation during the study with 1-hour interval monitoring.

undisturbed. In contrast, Barito Kuala in South Kalimantan is engaged in a different type of business: palm oil, where four palm oil corporations operate in the area and experiences shallower WTE due to the absence of companies engaged in the same business as Padang Island, which is pulp paper. The result showed that in Padang Island, WTE exceeding 40 cm occurs more than 80%, while in South

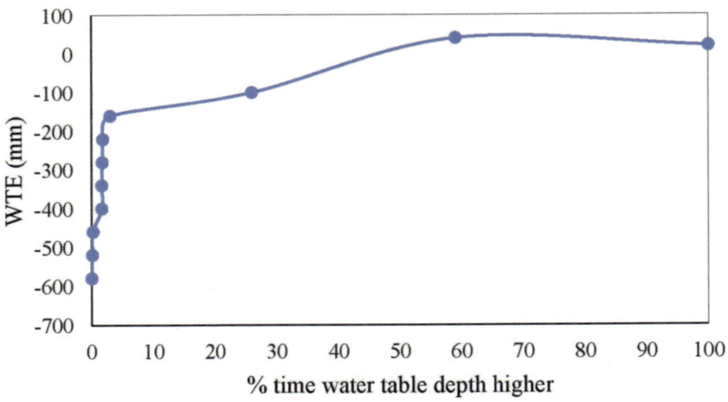

Figure 7.
Water table frequency curves of Jambu Gauge in Barito Kuala, South Kalimantan.

Kalimantan, it is less than 2%, and the interval in between reaches more than 70%. Nevertheless, besides the land use factor, other hydrology factors can contribute to this difference. Therefore, a comprehensive study needs to be conducted to reveal the cause of these hydrology changes.

Water table elevation can be used to observe its reaction to rainfall and flow. These phenomena can be seen in previous studies [10, 18]. They conducted their research in different peatlands, which are tropical and sub-tropical, yet they show the same phenomenon: the WTE responded to rainfall quite quickly, increasing as it generated flow. Both studies concluded that tropical and sub-tropical peatlands share the characteristic that the initial WTE and land moisture levels determine the responses of the WTE.

3.3 Parameters in peatlands and the interconnectivity

Hydrology parameters in wetlands, including peatlands, can serve as indicators of overall wetland conditions. Therefore, it would be interesting to understand how they reflect the peatland conditions. Hydrology parameters cover rainfall, water table elevations, evapotranspiration, surface and sub-surface flow, and flow hydrographs [2, 21–23]. Several approaches can be used to define the condition of peatlands. Some researchers define it by calculating a drought index to assess the likelihood of fire occurrence possibility in peatlands, while others describe it through time and spatial map water table elevation or the frequency of water table depth intervals [2, 24, 25]. Another way to illustrate the role of water table elevation is to explore its connection to vegetables [26–28]. There is no single definition for its role. *Sphagnum,* a natural vegetation, has been found to have a significant relationship with WTE [26, 29]. However, other plantations, such as *Acacia Crasicarpa*, which are human-planted, also exhibit this relationship without significant differences from natural forests. Furthermore, different vegetation remains the main indicator: water table elevation fluctuation [14, 30]. The relationship between rainfall and water table elevation can be observed across a wide range of scenarios, such as comparing monthly rainfall with maximum and average WTE values in the same month [2]. However, this relationship can manifest rapidly, as depicted in **Figure 8**.

As the data was expressed in a short period, the relationship can be seen in a very detailed time; it can be concluded that WTE responds to the rainfall range between

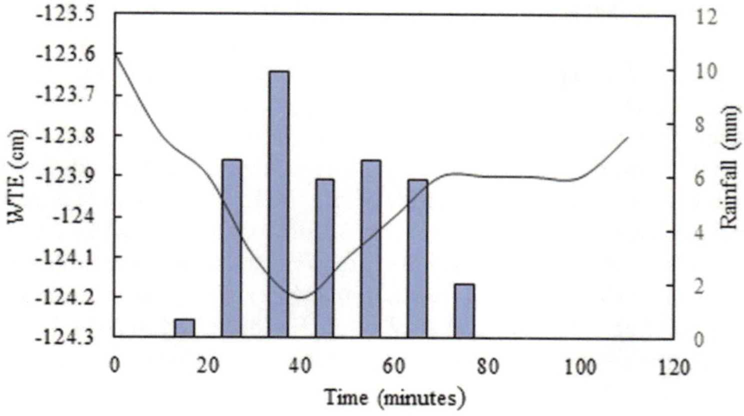

Figure 8.
Relationship between intensity and WTE in rainfall event on October 8, 2019, in Padang Island.

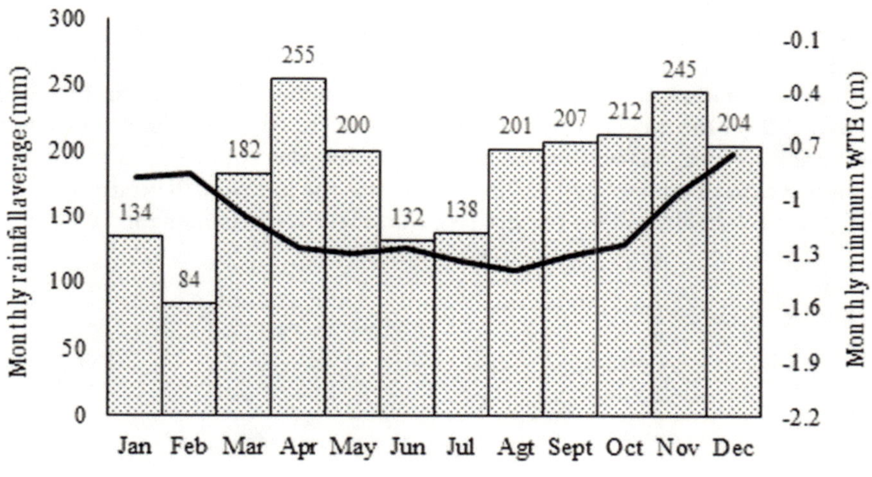

Figure 9.
Relationship between monthly rainfall average and monthly minimum WTE in Padang Island.

30 and 50 minutes to be constant, indicating significant fluctuations in the WTE during rainfall events. It becomes saturated in a very short time. This connection is not limited to presenting short-term data during specific rainfall events; it can also be observed every month, as illustrated in **Figure 9**.

There is a time difference between the occurrence of minimum rainfall in February and the minimum groundwater level in August, indicating the interval between when rain reaches its minimum and when groundwater level reaches its minimum. A similar phenomenon was observed in the previous research [31] conducted by Putra and Hayasaka [31] on the peatlands of Sebangau, Central Kalimantan, where an interval between the occurrence of minimum rainfall and minimum groundwater level elevation was noted, specifically in August and October. Still, the interval was smaller than the results of this research. The similarity is that there is an interval between the minimum rain and the minimum WTE, even though the gaps are different.

3.4 The relationship between water table elevation and fire hotspots

Another way to identify the relationship parameters can be presented by depicting the WTE distribution along the whole KHG and putting the hotspot area to analyze how much the WTE impacts fire in a peatland, as shown in **Figure 10**. WTE can be a sign of a drying situation in peatlands. Regarding the fire in 2014, the fire hotspot showed fire distribution in Padang Island, where the fire hotspot data is downloaded from the National Oceanic and Atmospheric Administration (NOAA) website. Based on the water table elevation variability, it is assumed that the lower table elevation (the furthest from the surface) will have a higher fire risk. The minimum WTE data are from the study between 2015 and 2017, while fire occurred in 2014. Nevertheless, the WTE can be assumed to show the general status of peatland dryness. The combined fire hotspots and the minimum WTE in Padang Island during the study can be presented in **Figure 10**.

Figure 10 shows that, in general, the fire happened in the expected area, but in a particular area, the fact was the reverse. On the northern island of Padang Island, the value of WTE was minimal, with more than 1 meter under the surface and the hotspots centered there. PPDs are the name for the location of water table elevation gauges. PPD 3 and PPD 4, which are adjacent to Sei Hiu, have been affected by a small fire hotspot, even though both are natural forests. On the other hand, in the middle of the island, PPD 8, the buffer zone with very high WTE, which is sometimes saturated above the soil surface, has no fire hotspots, as presumed.

A similar situation happened to PPD 9, which is the Industry Plantation area. As the WTE was relatively low, we can observe that there were no hotspots in PPD 9,

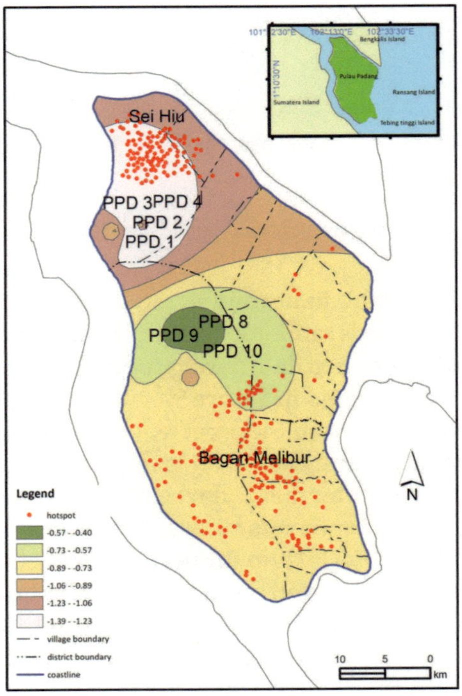

Figure 10.
Fire hotspot in 2014 in minimum WTE during the time study.

although it is the Industry Plantation area. Besides, it was laid next to the buffer zone, which had no fire hotspots. Unlike PPD 10, it is a natural forest with WTE, which is categorized the same as PPD 9. More explanation about how this PPD1 until PPD 10 related conditions can be found in a previous study [10].

4. Conclusion

Research in tropical peatlands required some specific approaches regarding their condition. This includes preparing practical equipment to be installed and gathering information about the local community. The connection with the community will determine the sustainability of the research, even if it may not seem directly related to the study. The type of equipment that can be installed depends on the available funding and the purpose of the data. If the study aims to explain the phenomenon in a term interval, it will require automatic data recording using a data logger at minute-level granularity. However, when attempting to describe the trends over daily or monthly intervals with limited funding, manual equipment that requires data to be recorded by an individual can be a viable alternative. Water table elevation (WTE) data, as a crucial parameter in peatlands, can be recorded at various intervals, including minutes, hours, days, and even months, depending on its intended use. WTE can also be an indicator of peatland conditions and the possibility of fire. Every interval data recording can be used for different purposes, so recognizing the study purposes and the limitations will make the research sustainable.

Author details

Nilna Amal[1*] and Didy Wurjanto[2]

1 Civil Engineering Department, Faculty of Engineering, Lambung Mangkurat University, Banjarbaru, Indonesia

2 Peatland and Mangrove Restoration Agency, Jakarta, Indonesia

*Address all correspondence to: nilna.amal@ulm.ac.id

References

[1] Vitt DH. Peatlands in Jorgensen SE, Fath BD, editors. Encyclopedia of Ecology. Elsevier B.V. 2008;**5**:2656-2664

[2] Amal N, Helda N, Rusdiansyah A, et al. Analysis of hydrology parameters in a tropical wetland as an early approach to identify a drought risk in a peatland area. IOP Conference Series: Earth and Environmental Science. 2022. DOI: 10.1088/1755-1315/999/1/012011

[3] Murdiyarso D, Lilleskov E, Kolka R. Tropical peatlands under siege: The need for evidence-based policies and strategies. Mitigation and Adaptation Strategies for Global Change. 2019;**24**(4):493-505

[4] Page SE, Siegert F, Rieley JO, Boehm HDV, Jaya A, Limin S. The amount of carbon released from peat and forest fires in Indonesia during 1997. Nature. 2002;**420**(6911):61-65

[5] Page S, Hosciło A, Wösten H, Jauhiainen J, Silvius M, Rieley J, et al. Restoration ecology of lowland tropical peatlands in Southeast Asia: Current knowledge and future research directions. Ecosystems. 2009;**12**(6):888-905

[6] Ohkubo S, Hirano T, Kusin K. Influence of fire and drainage on evapotranspiration in a degraded peat swamp forest in Central Kalimantan, Indonesia. Journal of Hydrology. 2021;**603**(PA):126906. DOI: 10.1016/j.jhydrol.2021.126906

[7] Putra EI. The effect of the precipitation pattern of the dry season on peat fire occurrence in the mega Rice project area, Central Kalimantan, Indonesia. Tropics. 2011;**19**(4):145-156

[8] Taufik M, Veldhuizen AA, Wösten JHM, van Lanen HAJ. Exploration of the importance of physical properties of Indonesian peatlands to assess critical groundwater table depths, associated drought and fire hazard. Geoderma. 2018;**2019**(347):160-169

[9] Ismail I, Haghighi AT, Marttila H, Kurniawan U, Karyanto O, Kløve B. Water table variations on different land use units in a drained tropical peatland island of Indonesia. Hydrology Research. 2021;**52**(6):1372-1388

[10] Amal N, Sujono J, Jayadi R, Ohgushi K. Variability of water table elevation and flow response of tropical peatland case study at Pulau Padang, Riau, Indonesia. Lowland Technology and International. 2021;**22**(4):135-141

[11] Susanti A, Karyanto O, Affianto A, Ismail PS, Aditya T, et al. Jurnal Ilmu Kehutanan. Jurnal Ilmu Kehutanan. 2017;**11**(1):29-42

[12] BRG-LKFT. Laporan Akhir Riset Pengelolaan Sumber Daya Air di Pulau Padang-Final Report Study of Water Resources in Padang Island. Yogyakarta, Indonesia: Universitas Gadjah Mada; 2017

[13] Arifin YF, Hamidah S, Fakhrurazi. Kajian Teknis Analisis Daya Dukung Hutan Lindung Kota Banjarbaru-Technical Study on Carrying Capacity Analysis of Banjarbaru City Protected Forests. Final Report. Lambung Mangkurat University; 2018

[14] Astiani D, Burhanuddin B, Curran LM, Mujiman M, Salim R. Effects of drainage ditches on water table level, soil conditions and tree growth of degraded peatland forests in

West Kalimantan. Indonesian Journal of Forestry Research. 2017;**4**(1):15-25

[15] Putra E, Cochrane MA, Vetrita Y, et al. Determining critical groundwater level to prevent degraded peatland from severe peat fire. IOP Conference Series: Earth and Environmental Science. IOP Publishing. 2018. DOI: 10.1088/1755-1315/149/1/012027

[16] Putra SS, Holden J, Baird AJ. The effects of ditch dams on water-level dynamics in tropical peatlands. Hydrological Processes. 2021;**35**:1-17

[17] Swails E, Yang X, Asefi S, Hergoualc'h K, Verchot L, RE MR, et al. Linking soil respiration and water table depth in tropical peatlands with remotely sensed changes in water storage from the gravity recovery and climate experiment. Mitigation and Adaptation Strategies for Global Change. 2019;**24**(4):575-590

[18] Daniels SM, Agnew CT, Allott TEH, Evans MG. Water table variability and runoff generation in an eroded peatland, south Pennines, UK. Journal of Hydrology. 2008;**361**(1-2):214-226. DOI: 10.1016/j.jhydrol.2008.07.042

[19] Ritzema H, Limin S, Kusin K, Jauhiainen J, Wösten H. Canal blocking strategies for hydrological restoration of degraded tropical peatlands in Central Kalimantan, Indonesia. Catena. 2014;**114**:11-20. DOI: 10.1016/j.catena.2013.10.009

[20] Amal N, Sujono J, Jayadi R. Water Table Variability and Flow Response of Tropical Peatland—A Case Study. In 3rd World Irrigation Forum (WIF3), Bali, Indonesia, 2019 International Commision on Irrigation and Drainage (ICID). 2019. pp. 1-10

[21] Schempf J. Wetland Classification, Inventory, and Assessment Methods: An Alaska Guide to their Fish and Wildlife Application, Technical Report. Alaska Department of Fish and Game. Habitat Division. 1992. Technical Report 93-2.

[22] Amal N. Analisis Karakteristik dan Formulasi Rawa dengan Pendekatan Variabel Hidrologi Rawa-Analysis of Wetland Characteristics and Formulation using the Wetland Hydrology Variable Approach. Info Teknik. 2021;**22**(1):99-116

[23] Hanson A, Swanson L, Ewing D, Grabas G, Meyer S, Ross L, et al. Wetland ecological functions assessment: An overview of approaches [Internet]. An Overview of Approaches. Canadian Wildlife Service Technical Report Series No. 497. Atlantic Region. 2008

[24] Novitasari N, Sujono J, Harto S, Maas A, Jayadi R. Drought index for peatland wildfire management in Central Kalimantan, Indonesia during el niño phenomenon. Journal of Disaster Research. 2019;**14**(7):939-948

[25] Wösten JHM, Clymans E, Page SE, Rieley JO, Limin SH. Peat-water interrelationships in a tropical peatland ecosystem in Southeast Asia. Catena. 2008;**73**(2):212-224

[26] Breeuwer A, Robroek BJM, Limpens J, Heijmans MMPD, Schouten MGC, Berendse F. Decreased summer water table depth affects peatland vegetation. Basic and Applied Ecology. 2009;**10**(4):330-339. DOI: 10.1016/j.baae.2008.05.005

[27] Haapalehto TO, Vasander H, Jauhiainen S, Tahvanainen T, Kotiaho JS. The effects of peatland restoration on water-table depth, elemental concentrations, and vegetation: 10 years of changes. Restoration Ecology. 2011;**19**(5):587-598

[28] Limpens J, Holmgren M, Jacobs CMJ, Van Der Zee SEATM, Karofeld E, Berendse F. How does tree density affect water loss of peatlands? A mesocosm experiment. PLoS One. 2014;**9**:1-11

[29] Kim J, Rochefort L, Hogue-Hugron S, Alqulaiti Z, Dunn C, Pouliot R, et al. Water table fluctuation in peatlands facilitates fungal proliferation, impedes sphagnum growth and accelerates decomposition. Frontiers in Earth Science. 2021;**8**(April):1-9

[30] Evans MG, Burt TP, Holden J, Adamson JK. Runoff generation and water table fluctuations in blanket peat: Evidence from UK data spanning the dry summer of 1995. Journal of Hydrology. 1999;**221**(3-4):141-160

[31] Putra EI, Hayasaka H. The effect of the precipitation pattern of the dry season on peat fire occurrence in the mega rice project area, central Kalimantan, Indonesia. Tropics; 2011;**19**:145-156

Chapter 3

A Novel Approach to Explore New Means of Depletion of Potable Water Crisis by Phytoremediation of Abandoned Coalmine Pitlake and Generate Alternate Livelihood: A Case Study of Raniganj Coalfield, West Bengal, India

Md Nazir, Kushal Roy, Ayan Saha and Dibyendu Saha

Abstract

Phytoremediation is using live plants to remove toxins and contaminants from land, water, or the air. Hydrocarbons, radionucleotides, fertilizers, explosives, heavy metals, and other contaminants are all efficiently removed from wastewater by this approach. Aquatic plants can be classified as free-floating, emergent, or submerged. Studies have been done on the capacity of Pontederia crassipes Mart. to extract pollutants like dissolved solids, and heavy metals from wastewater. The pollution of chromium in Indian chromite mining sites has also been addressed it and following phytoremediation, the biomass may be utilized to produce vermicompost, biogas, and bioethanol. Surface mining is producing changes in land use, land cover, and climate in the RCF area of West Bengal, India. By using aquatic plants to phytoremediate pit lakes (PLs), fresh water can be produced and chances for sustainable livelihoods can be created. The population in former mining regions has grown significantly, leading to the relinquishment of agriculture and the development of illegal mining and coal theft. This study aims to examine the literature on aquatic plant phytoremediation of water, emphasizing the necessity of phytoremediation of PL water in RCF and the potential applications of Pontederia crassipes Mart. biomass for the generation of biogas, bioethanol, and vermicompost.

Keywords: alternate livelihood, aquatic plants, phytoremediation, pit lake, Raniganj coalfields

IntechOpen

1. Introduction

Phytoremediation is characterized as the utilization of green plants and their corresponding microbes, soil enhancements, and the application of agronomic techniques to eliminate, contain, or render harmless environmental toxins [1, 2]. The three main categories of water pollutants are inorganic (such as metals or synthetic and manure-based fertilizers containing excess levels of N, and P), biological (such as pathogens and algal toxins), and organic pollutants including hydrocarbons, pesticides, and algal toxins [3]. Various aquatic plants have been evaluated for their phytoremediating ability explained by different authors globally. PLs are generated when dugouts made during surface-cut mining operations fill with water after the mining operations have been completed by groundwater recharge, surface water diversion, and monsoon precipitation over the years. PLs differ from natural lakes in that they have significantly higher relative depths. Today when the world is moving towards a water crisis, the abandoned PLs might act as an alternative supply of water because they contain enormous amounts of water. Regarding the phytoremediation of water in mining sites, numerous studies have been carried out in India and elsewhere in the world. The aquatic, and wetlands plant species are very effective at removing a wide range of organic, and inorganic pollutants from wastewater, including potentially toxic elements (PTE), fertilizers, explosives, radionucleotides, and hydrocarbons [4]. In this regard, it is worth mentioning that there are three categories into which aquatic plants can be divided free-floating, emergent, and submerged which can be used for effective phytoremediation [5]. After being purified, the water can be used sustainably for a variety of tasks, including hydroponics, irrigation, horticulture, aquaculture, etc. The phytoremediating ability of *Spirodela polyrhiza* (L.) Schleid. in the phytofiltration of As, and the mechanism of As uptake by the plant was studied in Dhaka, Bangladesh. For this, the plants were collected from a rice field in Manikgonj, Dhaka, Bangladesh, and were stock-cultured in a greenhouse using standard MS (Murashige and Skoog) media solution for 2 weeks [6]. The adsorption properties of a cationic dye (methylene blue, MB) on *Pontederia crassipes* Mart. in Egypt [7]. Likewise the effectiveness of *Pontederia crassipes* Mart. in removing Zn (II) and Cd (II) as well as their admixtures was assessed [8]. The effectiveness of *Pontederia crassipes* Mart. as a phytoremediation agent in the removal of metals from contaminated coastal areas of Nigeria. The evaluation was based on the presence of 10 metals in the water, including As, Cd, Cu, Cr, Fe, Mn, Ni, Pb, and Zn were evaluated from Ondo State in Nigeria's coastal region [9]. A Research was conducted on the ability of duckweed (*Lemna minor* L.) to break down the azo dye C.I. Acid Blue 92 (AB92) [10]. The removal of two harmful and cancer-causing contaminants, As and Cd, from water in Japan using an aquatic plant as *Micranthemum umbrosum* (J.F. Gmel.) S.F. Blake was investigated [11]. Whereas, in Chile, it was investigated how the aquatic plant *Oenothera picensis* Phil. extracted copper using compost and a biodegradable chelate called methylglycinediacetic acid [12]. *Scirpus grossus* L.f. is an aquatic plant, and it was investigated if this plant may assist Sago mill effluent (SME) in having reduced levels of TSS, COD, and BOD [13]. To determine the phytoremediating effectiveness of *Pistia stratiotes* L., the plant was exposed the plant to the physiological effects of artificial acid mine drainage (AMD) [14]. Looking at various research on the phytoremediation of aquatic weeds in India, the potential for uranium buildup, and subsequent biochemical reactions in *Hydrilla verticillata* (L.f.) Royle. was investigated [15]. A study showed that *Pontederia crassipes* Mart. is effective at removing dissolved solids, and heavy metals like Cr and Cu from wastewater, and investigated how the pH of wastewater changed as the grew in wastewater [16]. In the flying ash ponds of NTPC Unchahar, which is located in Uttar Pradesh, India, the potential of *Azolla caroliniana* Willd. as a heavy metal accumulator was investigated [17]. The effectiveness of *Pontederia crassipes* Mart. in the phytoremediation of Amravati River bank polluted dye industrial water was also studied [18]. *Azolla pinnata* R. Br. and *Lemna minor* L., two aquatic plants, were used in a study at the coal mines of Singrauli, Madhya Pradesh, India, to investigate the phytoremediation process for the detoxification of coal mine effluents [19]. A study was conducted in order to create a viable phytoremediation system for the efficient removal of hazardous hexavalent Cr from wastewater of the chromite mining sites in Sukinda, Orissa, India [20]. The aquatic species in that study to address the issue of Cr contamination in chromite mine wastewater was the *Pontederia crassipes* Mart [20]. After some biological processing, the biomass can be used for the manufacture of bioethanol, biogas, and vermicompost [21, 22]. In the RCF, West Bengal in India, surface mining, and its activities are the main force behind changes in land use, land cover, and climate. Since it opened 246 years ago, RCF, India the oldest coal mine, has been producing non-coking coal of the highest grade. The mining authority favors opencast mining over underground mining in order to lower the cost per unit of coal output [23]. PL clusters make up a significant percentage of RCF in West Bengal. Around 260 hectares of RCF, there are around 78 historic opencast coal pits [24, 25]. Over the years, the population has grown significantly in mining areas. Such townships are located above around 4400 million tons of significant coal reserves, which will have an impact on current and upcoming mining operations [26]. According to Guha [27], the extremely low rainfall and poor soil fertility have led to the abandonment of agriculture. Along with social ties, and organizational structures, human behavior has altered substantially. The locals no longer have any alternative means of support due to illegal mining and coal theft. They are unable to rely on fishing, agriculture, etc. [28]. They are compelled to work in the illegal industry as a quick method to get paid and as an alternative to subsistence farming. Nowadays it is happening so blatantly that there are often processions of misfit guys seen driving little vehicles and riding bicycles while hauling heavy loads of coal through the highways of Raniganj [28, 29]. Thus the phytoremediation of PLs by aquatic plants not only generates fresh water but also creates scopes for alternate livelihood through sustainable uses of PL water. The authors of this article are presently working on the phytoremediation of PLs in RCF by *Pontederia crassipes* Mart. and other aquatic plants, and using the biomass of the harvested aquatic plants in the production of biogas, bioethanol, and vermicompost with the noble approach of generating entrepreneuric scopes in the tribal-based regions.

The aim of the present study is to review various studies on the phytoremediation of water by aquatic plants, highlighting the need for phytoremediation of PL water in RCF, and using the phytoremediating biomass of *Pontederia crassipes* Mart. in the production of biogas, bioethanol, and vermicompost.

2. Methodology

2.1 Study site

The Asansol and Durgapur subdivisions of Paschim Bardhaman district in West Bengal, India are where the RCF is mostly encompassed with 23°37′44″N 87°06′54″

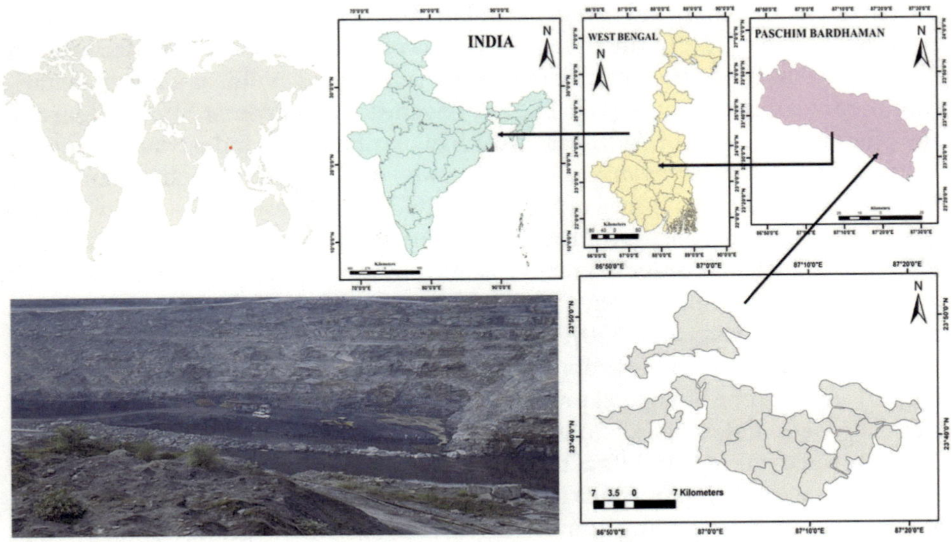

Figure 1.
Geographical location of RCF.

(**Figure 1**). According to Chattopadhyay [30], it extends to the Jharkhand districts of Dhanbad and its districts of Birbhum, Bankura, and Purulia in West Bengal, India. In RCF, coal mining initially got underway when coal was explored in 1774 by British East India Company employees John Sumner, and Suetonius Grant Heatly close to Ethora. Early exploration and mining operations were conducted in an indiscriminate way [30]. Alexander, and Co., a brokerage, spearheaded the beginnings of regular mining in 1820. With the purchase of the collieries by Prince Dwarkanath Tagore in 1835, Carr, Tagore, and Company dominated the industry. RCF served as the country's primary coal producer during the entirety of the 19th century and a significant portion of the 20th century [30]. The Coal Mines Authority of India was granted authority over all non-coking coal mines in 1973. Coal India Limited established Eastern Coalfields Limited in 1975 as a subsidiary. In RCF, it absorbed all of the preceding private collieries. The RCF has a total area of 443.50 km^2 and 49.17 billion tons of coal reserves in West Bengal and Jharkhand respectively [31]. In terms of reserves, the coalfield is the second-largest in the nation, 30.61 billion tons of the total reserve are in West Bengal, and 18.56 billion tons are in Jharkhand (**Figure 2**) [32].

2.2 Materials and methods

Drawing up the review employed a specific keyword-based hunt methodology. The keywords included phytoremediation by aquatic weeds, heavy metals, potable water crisis, freshwater status, coalmine pit lake, Raniganj Coalfield, etc. The chosen portals included Research Gate, Science Direct, PubMed, NCBI, and Google Scholar, among others. Results from the most recent three decades that is from 2001 to 2023 were approved for the custom search strategy.

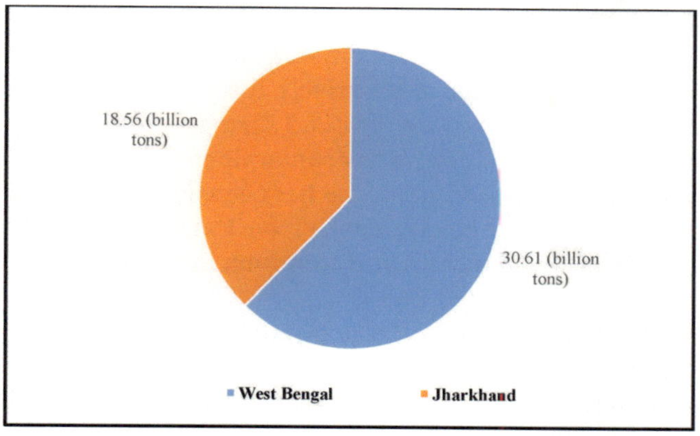

Figure 2.
Total coal reserve of RCF in West Bengal and Jharkhand [32].

3. Results and discussion

Azolla caroliniana Willd. and *Hydrilla verticillata* (L.f.) Royle, for their effectiveness in removing fly ash, and uranium, respectively, are the aquatic plants that have been mentioned as effective phytoremediation agents [15, 17]. Rahman et al. [6] reported on the efficiency of another species of duckweed, *Spirodela polyrhiza* (L.) Schleid, in removing As absorption through the phosphate uptake pathway. According to research on phytoremediation by Khataee et al. [10], duckweed plays a key role in the conversion of pollutants into intermediate molecules, and the elimination of dyes like Acid blue. *Oenothera picensis* Phil. and *Micranthemum umbrosum* (J.F. Gmel.) S.F. Blake are examples of aquatic plants that are effective at removing As, and Cd, respectively, according to recent studies by Gonzalez et al. [12] and Islam et al. [11]. According to Bharti et al. [19], both the plants, *Azolla pinnata* R. Br. and *Lemna minor* L., are suitable for phytoremediation of mine effluents at contaminated locations. It was also claimed that due to the high level of pollutants in these phytoremediated plants, it is necessary to pay attention to developing disposal plans [19]. Novita et al. [14] showed that *Pistia stratiotes* L. can change the artificial AMD's pH level from 4 to 7.3 in just 14 days. With initial concentrations of 5.3 mg/l and 7.5 mg/l, respectively, this plant can also reduce the copper content by up to 92.45% and 88.00% in 14 days, with the highest removal occurring on day 3. It is clear that *Pistia stratiotes* L. has the power to lower copper levels and neutralize pH in artificial AMD. It was claimed that the SS batch system outperforms the FS batch system for the survival of the plant *Scirpus grossus* L.f. Additionally, the plant has the ability to reduce TSS, COD, and BOD by 98.88, and 93%, respectively [13]. We can therefore draw the conclusion that this plant has extraordinary potential for removing contaminants during the phytoremediation of SME. According to Chandra and Yadav [33], *Typha angustifolia* L. has the ability to control Cd, Cr, Cu, Fe, Mn, Ni, and Pb levels in water. The phytoremediation process in abandoned mines in Korea is currently under evaluation. In this context, Lee et al. [34] discussed the difficulties with this phytoremediation approach and asserted that additional study is necessary to comprehend the interactions between plant roots, microbes, the soil, and contaminants, that make up the rhizosphere, in order to make advances in the field of phytoremediation of Korea.

Khaiary et al. [7] discussed the widespread usage of *Pontederia crassipes* Mart. for phytoremediation of dye-contaminated water. According to Hasan et al. [8], *Pontederia crassipes* Mart. is capable of effectively absorbing Zn (II) and Cd (II) from aqueous solutions up to concentrations of 6, and 2.5 mg/L, respectively. The effectiveness of the plant in phytoremediating water contaminated by PTEs like Cd, As, Pb, Cr, microbial toxins, and suspended particles has been documented. The plant was also able to remove 99.5% of the hexavalent Cr from the wastewater. Additionally, this purifying system is both economically, and environmentally sound and does not need much expert opinion [9, 16, 18, 20].

The fuel crisis is a major issue that the globe is currently experiencing. Moreover, air pollution is a result of the burning of fuels. The use of ethanol, a green energy, as a fuel substitute in this scenario has been studied by various authors which also limits air pollution. However, ethanol production is economically challenging so bioethanol must be considered in this regard. The phytoremediated biomass of water hyacinth is a rich source of lignocellulose which can be further reused to produce bioethanol. Pretreatment, hydrolysis, and fermentation are the three steps in the synthesis of bioethanol (**Figure 3**) [35]. Upon pretreatment, the lignocellulosic biomass produces lignin, and fermentable carbohydrates [36–38]. By making more cellulose, and hemicellulose accessible for the hydrolysis stage, this process removes lignin and modifies the structure of cellulose and hemicellulose [39].

According to Lin et al. [40], there are numerous approaches to pretreating lignocellulosic biomass, including mechanical pretreatment, alkali or acid pretreatment, and steam explosion. The use of biological agents like fungi as part of other pretreatment techniques is known as biological pretreatment. Commonly utilized as biological pretreatment agents include white-rot fungi, brown-rot fungi, and soft-rot fungi [21, 22, 41]. After pretreatment, cellulose, and hemicellulose are hydrolyzed to yield straightforward fermentable sugars. It is more difficult to hydrolyze cellulose than hemicellulose because of its crystalline structure. Because of this, cellulose hydrolysis is always performed with the aid of an acid or specialist enzyme (cellulase). The breakdown of cellulose is referred to as biochemical hydrolysis or saccharification, whereas the hydrolysis of hemicelluloses is referred to as chemical or enzymatic hydrolysis [42]. Following hydrolysis, the biological process of fermentation takes place. Which eventually reduces simple carbohydrates to smaller molecules like acids, and alcohols. For this stage, microorganisms are required. To catalyze chemical reactions during fermentation, these bacteria produce enzymes. Despite

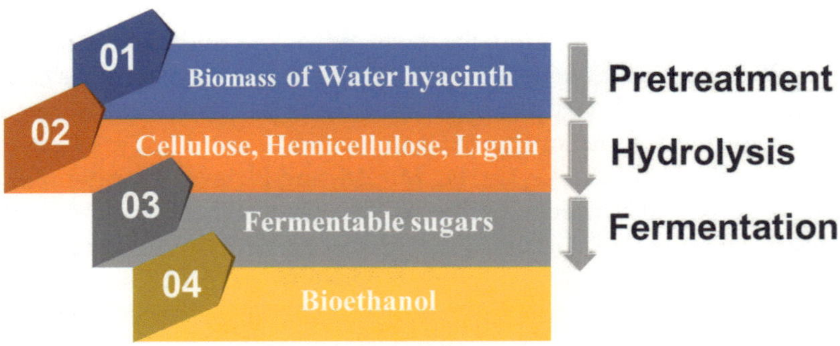

Figure 3.
Flowchart showing bioethanol preparation [35].

the fact that these sugars are essential for the metabolism, and reproduction of all living things, a number of bacteria and yeasts are capable of fermenting them. The two most popular species employed to produce ethanol from hexose sugars on a big scale are *Saccharomyces cerevisiae* and *Zymomonas mobilis*. *Saccharomyces cerevisiae* is the species that produces bioethanol most frequently. The durability, and suitability of this strain of yeast for the fermentation of glucose from lignocellulosic biomass are its main advantages. The gram-negative, facultatively anaerobic species of bacterium *Zymomonas mobilis*, on the other hand, is less frequently employed for the fermentation process due to its lower yield and more active metabolism, which makes it more prone to contamination [43–45]. The anaerobic digestion of *Pontederia crassipes* Mart. to produce biogas is a practical alternative strategy for the provision of environmentally friendly energy and aquatic weed management. The presence of firmly bound refractory lignin, crystalline cellulose, and hemicellulose polymers slows anaerobic digestion and restricts the production of biogas [46]. If collected biomass is valuable for use in converting biomass or other product preparation, aquatic restoration might move along more swiftly. It serves as a substrate for the breakdown of organic matter or the production of biogas, which can yield fertilizer-making byproducts from digestion. According to past research on this subject by Priya et al. [47], *Pontederia crassipes* Mart. can contain a lot of nitrogen, up to 3.2% of its dry mass, and have a C/N ratio of about 15. The inoculant sludge was made using the wastewater from a biogas system that processed cow manure. The ratio of water to *Pontederia crassipes* Mart. used to make the slurry was 1:10 (w/w). The slurry underwent H_2SO_4 pretreatment. The slurry's volatile solid content was increased by the addition of water until it reached 4% g/g Total Mass of Slurry (TMS). Using NaOH, the pH of the slurry was kept between 6 and 8. The active sludge was applied using an inoculum ratio of slurry: sludge of 75:25 with a total volume of 600 ml. Pretreating slurry with H_2SO_4 at a concentration of 5% v/v for 60 minutes produced the highest methane content (64.38%) and the maximum amount of total biogas (424.30 ml) (**Figure 4**) [48]. NPK concentration in vermicompost is higher than in the original feed supply. The vermicompost C/N ratio rose as a result of earthworm activity. Earthworm activity increased bacteria, fungi, and actinomycetes in the microflora. Additionally, they deliver the nutrients required for the growth of the microbes. In vermicompost, which has higher enzyme activity than typical compost earthworms and an aerobic heterotrophic microbial community drive the breakdown process (**Figure 5**) [37, 49].

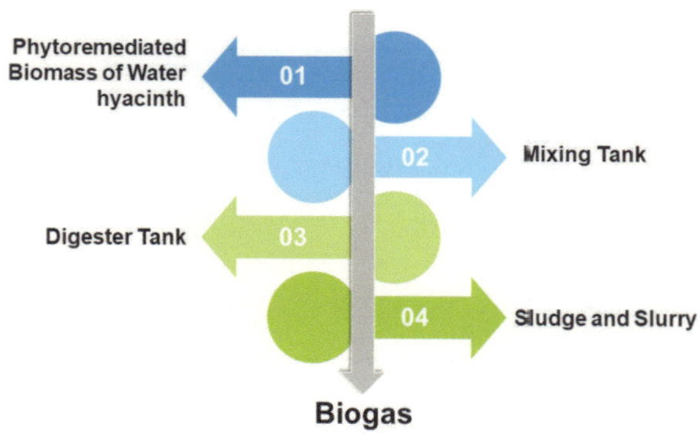

Figure 4.
Preparation of biogas [48].

Figure 5.
Vermicompost production [49, 37].

Water is a necessary component for life to exist on Earth. Water is required for both life on Earth and nearly all other human endeavors. As a result, water is necessary for all life. Our planet is around 75% covered in water. The total volume of water on Earth is approximately 1.386 billion km^3, of which 97.5% is salt water. Eakins et al. [50] estimated the total oceanic volume to be 1332 x 10^6 km^3. Therefore, freshwater (FW) makes up only 2.5% of water (**Figure 6**). Furthermore, just 0.3% of the total water supply is surface water, and the great majority of freshwater is deposited as glaciers [51, 52]. UNICEF estimates that nearly 4 billion people, or nearly 65% of the world's population, experience extreme water scarcity for at least one month; nearly 2 billion people live in areas with inadequate water supplies; by 2025, nearly half of the world's population will experience water scarcity; by 2030, 700 million people will have to relocate due to severe water constraints; and by 2040, 25% of children will have to live in areas with extreme water stress. The lack of freshwater resources worldwide is a significant barrier to advancing human progress and realizing sustainable development objectives [53]. More people are realizing that the FW shortage is a widespread structural issue on a worldwide scale [54]. Between 0.5 and 3.1 billion people will experience greater water scarcity by the year 2050 as a result of climate change, according to a study by Gosling et al. [55]. His study made use of climatic predictions based on numerous global climate models [55].

With an average annual rainfall of 1150–1450 mm, the RCF area receives 80–90% of its rainfall during the monsoon season. The Damodar, Ajay, and Barakar rivers govern the drainage system.

According to several studies [56–58], the geological structure of the RCF region, where water table depths typically range from 3 to 15 meters below ground level,

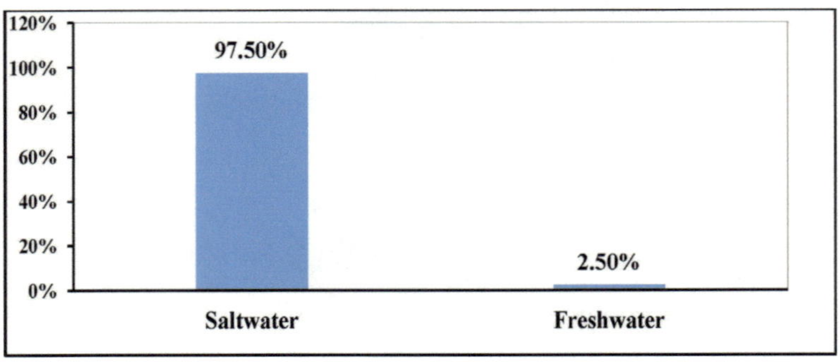

Figure 6.
Percentage of saltwater and freshwater according to Kundzewicz et al. [51]; Oki et al. [52].

Figure 7.
An abandoned PL in RCF.

significantly affects the occurrence and storage of groundwater. In this scenario, PL water (**Figure 7**) can be an alternate source of FW following purification in accordance with the Jal Jeevan Mission and Swachh Bharat Mission of the Government of India.

The purified water can be used sustainably in agriculture. One of the most cutting-edge methods for utilizing abandoned PL sustainably is hydroponics, which succeeded in phytoremediation. In hydrophonic farming, instead of using soil to grow the plants, nutrient-based water solutions are used. According to Helmreich [59], Dhananjani, and Pakeerathan [60], this kind of farming needs some fundamental conditions, such as access to clean water, oxygen, nutrients, and light, as well as root support materials including peat moss, vermiculite, perlite, and coconut fiber. The area where plants have been grown can be supplied with nutrient-rich water via a pump. The pots must be chosen so that the roots may stay submerged in water. As the plants absorb the nutrients from the water, the nutrient-rich water is provided at regular intervals. It is emptied of the extra water. Moreover, other ways of sustainability may include horticulture, and pisciculture In order to produce bioethanol, biogas, and vermicompost, the *Pontederia crassipes* Mart. biomass after phytoremediation can be employed. As a result, multiple levels of entrepreneurship will be possible. Bioethanol manufacturing processes can employ the gathered biomass as a raw material. Additionally, a number of companies can be established at the cooperative and small-scale levels to market bioethanol, biogas, and vermicompost. The production of vermicompost can also support the development of agriculture in that region. Thus our approach also supports the Make in India initiative by the Government of India according to which opportunities for new sources of income can be created as alternate livelihoods for the parishioners, who are otherwise involved in illegal practices.

4. Conclusion

There are effective mine-closing procedures in wealthy nations. However, developing countries like India observe major challenges in the implementation of such programmes. In addition to revealing alternative means of subsistence in RCF regions, the authors attempt to propose sustainable mine closure plans for surface-cut mining. The initial part of the plan is to phytoremediate PL water, making PLs an alternate supply of fresh water. Additionally, the authors propose the commercialization of bioethanol, biogas, and vermicompost made from the harvested biomass of phytoremediated aquatic plants. Furthermore, in order to address the major challenges of the water crisis, food insecurity, and entrepreneurship, the authors are currently working on the development of pollutant-resistant crop varieties to encourage agriculture in polluted soil of the catchment basin area of coal mine PLs.

Acknowledgements

The authors are grateful to the head of the botany department at the University of Burdwan in West Bengal, India.

Conflict of interest

The authors declare that they have no conflict of interest.

Author details

Md Nazir[1], Kushal Roy[2], Ayan Saha[1] and Dibyendu Saha[1*]

1 Department of Botany, The University of Burdwan, West Bengal, India

2 Department of Botany, Ramakrishna Mission Vivekananda Centenary College, West Bengal State University, West Bengal, India

*Address all correspondence to: dsbotbu79@gmail.com

References

[1] Das PK. Phytoremediation and nanoremediation: Emerging techniques for treatment of acid mine drainage water. Defence Life Science Journal. 2018;**3**(2):190-196

[2] Costello C. Acid mine drainage: innovative treatment technologies. In: US Environmental Protection Agency, Office of Solid Waste and Emergency Response. Washington, DC: Technology Innovation Office; 2003

[3] Fletcher J, Willby N, Oliver D, Quilliam R. Phytoremediation using aquatic plants. In: Shmaefsky B, editor. Phytoremediation Concepts and Strategies in Plant Sciences. Springer; 2020. pp. 205-260

[4] Dhir B. Phytoremediation: Role of Aquatic Plants in Environmental Clean-Up. 1st ed. Vol. 14. New Delhi: Springer; 2013. pp. 1-111

[5] Ryder S. The Embassy of the Mantenhas. Johannesburg: University of Johannesburg (South Africa); 2016. pp. 1-45

[6] Rahman MA, Hasegawa H, Ueda K, Maki T, Okumura C, Rahman MM. Arsenic accumulation in duckweed (*Spirodela polyrhiza* L.): A good option for phytoremediation. Chemosphere. 2007;**69**(3):493-499

[7] El-Khaiary MI. Kinetics and mechanism of adsorption of methylene blue from aqueous solution by nitric-acid treated water-hyacinth. Journal of Hazardous Materials. 2007;**147**(1-2):28-36

[8] Hasan SH, Talat M, Rai S. Sorption of cadmium and zinc from aqueous solutions by water hyacinth (Eichchornia crassipes). Bioresource Technology. 2007;**98**(4):918-928

[9] Agunbiade FO, Olu-Owolabi BI, Adebowale KO. Phytoremediation potential of *Eichornia crassipes* in metal-contaminated coastal water. Bioresource Technology. 2009;**100**(19):4521-4526

[10] Khataee AR, Movafeghi A, Torbati S, Lisar SS, Zarei M. Phytoremediation potential of duckweed (*Lemna minor* L.) in degradation of CI acid blue 92: Artificial neural network modeling. Ecotoxicology and Environmental Safety. 2012;**80**:291-298

[11] Islam MS, Saito T, Kurasaki M. Phytofiltration of arsenic and cadmium by using an aquatic plant, *Micranthemum umbrosum*: Phytotoxicity, uptake kinetics, and mechanism. Ecotoxicology and Environmental Safety. 2015;**112**:193-200

[12] González I, Neaman A, Cortés A, Rubio P. Effect of compost and biodegradable chelate addition on phytoextraction of copper by *Oenothera picensis* grown in Cu-contaminated acid soils. Chemosphere. 2014;**95**:111-115

[13] Nash DAH, Abdullah SRS, Hasan HA, Idris M, Othman AR, Al-Baldawi IA, et al. Utilisation of an aquatic plant (*Scirpus grossus*) for phytoremediation of real sago mill effluent. Environmental Technology & Innovation. 2020;**19**:1-11

[14] Novita VZ, Moersidik SS, Priadi CR. Phytoremediation potential of *Pistia stratiotes*to reduce high concentration of copper (Cu) in acid mine drainage. IOP Conference Series: Earth and Environmental Science. 2019;**355**(1):012063

[15] Srivastava S, Bhainsa KC, D'Souza SF. Investigation of uranium accumulation potential and biochemical responses of an aquatic weed *Hydrilla verticillata*

(lf) Royle. Bioresource Technology. 2010;**101**(8):2573-2579

[16] PNAML, Madhu G. Removal of heavy metals from waste water using water hyacinth. International journal on transportation and Urban Development. 2011;**1**(1):48-52

[17] Pandey VC. Phytoremediation of heavy metals from fly ash pond by *Azolla caroliniana*. Ecotoxicology and Environmental Safety. 2012;**82**:8-12

[18] Panneerselvam B, Priya K, S. Phytoremediation potential of water hyacinth in heavy metal removal in chromium and lead contaminated water. International Journal of Environmental Analytical Chemistry. 2023;**103**(13):3081-3096

[19] Bharti S, Banerjee TK. Phytoremediation of the coalmine effluent. Ecotoxicology and Environmental Safety. 2012;**81**:36-42

[20] Saha P, Shinde O, Sarkar S. Phytoremediation of industrial mines wastewater using water hyacinth. International Journal of Phytoremediation. 2017;**19**(1):87-96

[21] Ayanda OI, Ajayi T, Asuwaju FP. Eichhornia crassipes (Mart.) Solms: Uses, challenges, threats, and prospects. The Scientific World Journal. 2020;**2020**:1-12

[22] Bote MA, Naik VR, Jagadeeshgouda KB. Review on water hyacinth weed as a potential bio fuel crop to meet collective energy needs. Materials Science for Energy Technologies. 2020;**3**:397-406

[23] Patra T, Dutta D, Kundu A, Kumar M, Hossain SS, Chattoraj KK. Evolution of opencast mines in the raniganj coalfield (India): An assessment through multi-temporal satellite data. Journal of the Geological Society of India. 2022;**98**(3):387-394

[24] Kelkar Khambete. The Pit Lakes of Raniganj. 2023. Retrieved September 2, 2023. Available from: https://www.indiawaterportal.org/articles/pit-lakes-raniganj

[25] Saha D, Keshri JP, Saha NC. Comprehensive study on Raniganj coalfield area, India: A review. Ecology Environment & Conservation. 2022;**28**:S387-S398

[26] Gupta AK, Dutta AK, Basu R. Subsidence—A major effect of coal mining in Raniganj coalfield. IJRDO Journal of Business Management. 2018;**4**(8):14-27

[27] Guha D. A case study on the effects of coal mining in the environment particularly in relation to soil, water and air causing a socio-economic Hazard in Asansol-Raniganj area, India. International Journal of Social Science. 2014;**3**(8):39-42

[28] Tiwari I. The impact of environmental litigation in India: With reference to illegal mining. Issue 1 Indian Journal of Law and Legal Research. 2023;**5**(1):1

[29] Saha S. Illegal Mining Forces Raniganj Residents to Eat Now and Die Tomorrow. 2019. Retrieved September 2, 2023. Available from: https://www.downtoearth.org.in/news/general-elections-2019/illegal-mining-forces-raniganj-residents-to-eat-now-and-die-tomorrow-64206

[30] Chattopadhyay. Bardhaman Jelar Itihas O Lok Sanskriti. Radical. 2001;**1**:46-51

[31] Eastern Coalfields Limited. geologydata.info. Retrieved September

2, 2023. Available from: http://ww1.geologydata.info/?subid1=7af40e9e-49aa-11ee-a6e7-57bab2d4fcf1

[32] Overview of ECL Production & Productivity. https://www.easterncoal.nic.in/. Retrieved September 2, 2023. Available from: http://secureloginecl.co.in/elearning/contents/OVER%20VIEW%20OF%20ECL.pdf

[33] Chandra R, Yadav S. Phytoremediation of Cd, Cr, Cu, Mn, Fe, Ni, Pb and Zn from aqueous solution using *Phragmites cummunis, Typha angustifolia and Cyperus esculentus*. International Journal of Phytoremediation. 2011;**13**(6):580-591

[34] Lee SH, Park H, Kim JG. Current status of and challenges for phytoremediation as a sustainable environmental management plan for abandoned mine areas in Korea. Sustainability. 2023;**15**(3):2761. 1-2761.13

[35] Dave N, Selvaraj R, Varadavenkatesan T, Vinayagam R. A critical review on production of bioethanol from macroalgal biomass. Algal Research. 2019;**42**(101606):1-14

[36] Mankar AR, Pandey A, Modak A, Pant KK. Pretreatment of lignocellulosic biomass: A review on recent advances. Bioresource Technology. 2021;**334**:125235

[37] Alami E, Karimi M, Chalavi V. Investigation of compost and vermicompost of wate hyacinth as growing media for lily (Longiflorum× Asiatic). International Journal of Horticultural Science and Technology. 2021;**8**(3):271-280

[38] Singh S, Cheng G, Sathitsuksanoh N, Wu D, Varanasi P, George A, et al. Comparison of different biomass pretreatment techniques and their impact on chemistry and structure. Frontiers in Energy Research. 2015;**2**(62):1-12

[39] Sun Y, Cheng J. Hydrolysis of lignocellulosic materials for ethanol production: A review. Bioresource Technology. 2002;**83**(1):1-11

[40] Lin R, Cheng J, Song W, Ding L, Xie B, Zhou J, et al. Characterisation of water hyacinth with microwave-heated alkali pretreatment for enhanced enzymatic digestibility and hydrogen/methane fermentation. Bioresource Technology. 2015;**182**:1-7

[41] Kaur M, Gill MK, Sharma S, Kocher GS, Sodhi HS. Biological Pretreatment strategies for second-generation Lignocellulosic biomass to enhance ethanol production. In: Agroindustrial Waste for Green Fuel Application. Springer Nature Singapore: Singapore; 2023. pp. 169-203

[42] Houfani AA, Anders N, Spiess AC, Baldrian P, Benallaoua S. Insights from enzymatic degradation of cellulose and hemicellulose to fermentable sugars–a review. Biomass and Bioenergy. 2020;**134**(105481):1-13

[43] Xia J, Yang Y, Liu CG, Yang S, Bai FW. Engineering *Zymomonas mobilis* for robust cellulosic ethanol production. Trends in Biotechnology. 2019;**37**(9):960-972

[44] Liu CG, Xiao Y, Xia XX, Zhao XQ, Peng L, Srinophakun P, et al. Cellulosic ethanol production: Progress, challenges and strategies for solutions. Biotechnology Advances. 2019;**37**(3):491-504

[45] Rastogi M, Shrivastava S. Recent advances in second generation bioethanol production: An insight to pretreatment, saccharification and fermentation

processes. Renewable and Sustainable Energy Reviews. 2017;**80**:330-340

[46] Barua VB, Goud VV, Kalamdhad AS. Microbial pretreatment of water hyacinth for enhanced hydrolysis followed by biogas production. Renewable Energy. 2018;**126**:21-29

[47] Priya P, Nikhitha SO, Anand C, Nath RD, Krishnakumar B. Biomethanation of water hyacinth biomass. Bioresource Technology. 2018;**255**:288-292

[48] Sarto S, Hildayati R, Syaichurrozi I. Effect of chemical pretreatment using sulfuric acid on biogas production from water hyacinth and kinetics. Renewable Energy. 2019;**132**:335-350

[49] Sridevi S, Prabu M, Tamilselvi NG. Bioconversion of water hyacinth into enriched vermicompost and its effect on growth and yield of peanut. International Journal of Current Microbiology and Applied Sciences. 2016;**5**(9):675-681

[50] Eakins BW, Sharman GF. Volumes of the World's oceans from ETOPO2v2. AGU Fall Meeting Abstracts. 2007;**2007**:OS13A-O0999

[51] Kundzewicz ZW, Mata LJ, Arnell NW, Doll P, Kabat P, Jimenez B, et al. Freshwater resources and their management. In: Parry ML, Canziani OF, Palutikof JP, van der Linden PJ, Hanson CE, editors. Climate Change 2007: Impacts, Adaptation and Vulnerability. Contribution of Working Group II to the Fourth Assessment Report of the Intergovernmental Panel on Climate Change. New York: Cambridge University Press; 2007. pp. 173-210. ISBN 9780521880091

[52] Oki T, Kanae S. Global hydrological cycles and world water resources. Science. 2006;**313**(5790):1068-1072

[53] Dolan F, Lamontagne J, Link R, Hejazi M, Reed P, Edmonds J. Evaluating the economic impact of water scarcity in a changing world. Nature Communications. 2021;**12**(1):1-10

[54] Mekonnen MM, Hoekstra AY. Four billion people facing severe water scarcity. Science Advances. 2016;**2**(2):e1500323

[55] Gosling SN, Arnell NW. A global assessment of the impact of climate change on water scarcity. Climatic Change. 2016;**134**:371-385

[56] Manna A, Maiti R. Impact of open-pit mining on surface hydrology in Raniganj mining area, India. Mine Water and the Environment. 2016;**35**:156-167

[57] Banerji S, Mitra D. Geographical information system-based groundwater quality index assessment of northern part of Kolkata, India for drinking purpose. Geocarto International. 2019;**34**(9):943-958

[58] Kumar P, Singh AK. Hydrogeochemistry and quality assessment of surface and sub-surface water resources in Raniganj coalfield area, Damodar Valley, India. International Journal of Environmental Analytical Chemistry. 2022;**102**(19):8346-8369

[59] Helmreich S. Hydrophonic fields. In: Bull M, Cobussen M, editors. The Bloomsbury Handbook of Sonic Methodologies. New York: Bloomsbury Publishing; 2020. pp. 1-480

[60] Dhananjani BA, Pakeerathan K. Organic nutrient solutions for Hydrophonic spinach (*Basella alba*) production in urban agriculture. Journal of Agricultural Sciences—Sri Lanka. 2023;**8**(1):1-13

Chapter 4

Integrated Modeling of Hydrosedimentological Processes for Port Resilience and Environmental Assessment for Decision-Making: Case Study of the Paraíba do Sul River, Brazil

Lidiane dos Santos Lima, Paulo Cesar Colonna Rosman and Juliane Castro Carneiro

Abstract

The chapter evaluates the importance of integrated hydrosedimentological modeling in improving port resilience and environmental decision-making under the context of climate change, using the Paraíba do Sul River, Brazil, as a case study. It highlights the use of advanced predictive modeling, such as SWAT and SisBaHiA, in addressing the challenges posed by extreme climatic events on port operations and water resource management. The research emphasizes the need for tools to anticipate and adapt to adverse impacts, integrating remote sensing technologies and GIS for precise environmental monitoring. The goal is to promote informed decision-making for the sustainable management of water resources, mitigating erosion and sedimentation and adapting port operations to climate change.

Keywords: hydrosedimentological modeling, port resilience, climate change, remote sensing, water resource management, SWAT, SisBaHiA

1. Introduction

In the global context of increasing frequency of extreme weather events, port areas face unprecedented challenges. The complexity of sediment processes and water flows, intensified by such events, demands innovative approaches to ensure the operational continuity and sustainability of ports. This study focuses on the application of integrated hydrosedimentological modeling in the Paraíba do Sul River region, Brazil, aiming to formulate strategies that strengthen the resilience of port infrastructures and improve water resource management in the face of challenges posed by climate change [1, 2].

The main question guiding this study is: "How can integrated predictive modeling of hydrological and sedimentological processes be improved to effectively predict and mitigate these impacts, ensuring the resilience and sustainability of port operations in the face of climate variations?" This question reflects the urgency to develop advanced predictive modeling capable of predicting and mitigating the adverse effects of extreme precipitation events, underlining the need to adapt port operations to climate change expectations [3, 4].

The analysis and prediction of the effects of climate change on port areas, through hydrological and sedimentological modeling studies, are crucial to ensure the sustainability and resilience of these vital infrastructures [5, 6]. Instruments such as the SWAT (Soil and Water Assessment Tool) and the SisBaHiA (Environmental Hydrodynamics Base System) are fundamental in this area, providing a thorough assessment of sedimentation and erosion processes, as well as their impacts on port activities. These modeling tools offer essential insights into the dynamic mechanisms that influence coastal and port areas, enabling the development of proactive strategies for environmental management and the optimization of port operations in the face of challenges posed by climate change [7, 8].

The SWAT (Soil and Water Assessment Tool) is a tool used to analyze the impact of changes in land use and management practices on hydrology and sediment transport [9]. The combination of SWAT with remote sensing technologies and Geographic Information Systems (GIS) has been shown to be particularly effective in Land Use and Cover Change (LUCC) studies, emphasizing the efficiency of these technologies in monitoring and predicting environmental changes with high accuracy [10] illustrate how this integration can provide detailed and robust insights into environmental dynamics, facilitating informed decision-making for the sustainable management of water resources and the mitigation of erosion and sedimentation impacts.

SisBaHiA (Base System of Environmental Hydrodynamics) is a tool focused on hydrosedimentological and water quality modeling in estuarine and coastal environments. Its use is essential to obtain understandings about the influence of currents, tides, and sedimentation processes on the configuration of port areas. Through SisBaHiA, it is possible to identify how port morphology is affected by these natural factors and develop effective mitigation strategies against erosion and siltation. The application of this model allows the development of specific solutions that aim to preserve the functionality and sustainability of port infrastructures, ensuring their operability in the face of challenges imposed by coastal dynamics and sedimentological processes [11–13].

Climate change presents notable challenges to port operations, affecting precipitation patterns, sea levels, and the occurrence of extreme weather events. Research focusing on the protective effects of coral reefs, seagrass meadows, and mangroves underlines the vital importance of maintaining healthy coastal ecosystems. These ecosystems play a crucial role in defending against erosion and providing valuable ecosystem services, contributing significantly to the protection of port areas from adverse environmental impacts [14, 15].

In addition, assessing soil quality and understanding its influence on sediment transport are important for the development of effective mitigation strategies. Using soil quality indices to compare different methods offers an objective way to assess soil health and its ability to sustain vegetation, a crucial element in preventing erosion. This approach allows for a better understanding of how land management practices can be optimized to protect coastal and port areas from degradation [16].

The challenge lies in the effective adaptation of integrated predictive modeling approaches to capture the complexity and variability of climate impacts, both on a local and global scale, in port zones. The main gap lies in the definition of specific methodologies and the implementation of high-resolution data that allow an accurate prediction of the interaction between extreme precipitation events and sedimentary and water dynamics, taking into account future climate variations [17, 18].

Integrated predictive modeling is essential for water resource management, urban planning, climate change adaptation, and disaster mitigation. It combines several models and techniques to predict the impact of environmental, social, and economic variables on complex systems. Integrating hydrological models such as SWAT with water quality models, such as SisBaHiA, and land use models provides a comprehensive understanding of the processes studied. This approach is vital for addressing today's environmental, social, and economic challenges, offering a powerful tool for complex systems analysis, accurate predictions, and informed decision-making at various scales, enabling the development of sustainable and resilient solutions [19].

This work aims to improve the understanding and practices in integrated predictive modeling to predict and mitigate impacts of extreme precipitation events in ports, taking into account meteorological, hydrological, sedimentological, oceanographic, and geographic variations. Thus, a new modeling methodology is proposed to increase port resilience and sustainability against climate change. The research seeks to advance knowledge in hydrology, offering practical recommendations to strengthen port operations in the face of the climate challenge.

The gap highlighted, emphasizing the need for refinement in modeling methodologies and understanding of interactions in hydrological and sedimentological systems during such extreme events, aligns with the goal of this research. Improving predictive modeling to predict and minimize impacts requires an in-depth understanding of these dynamics, underlining the importance of exploiting these gaps in studies with the use of advanced technologies and analysis under different climatic conditions.

Conducting this research offers significant benefits, including enhancing the ability to predict and mitigate risks associated with extreme weather events in ports. This facilitates the development of more effective strategies to manage water and sedimentary resources, increasing the resilience of port infrastructures. In addition, it contributes to the environmental and economic sustainability of port operations by informing public policies and management practices that are adaptive to climate change.

2. Methodology

The study explored the integrated modeling of hydrological and sedimentological processes in Brazil in the Paraíba do Sul and Itabapoana River Basins, as well as in the adjacent coastal zone, using two modeling systems: the Base System of Environmental Hydrodynamics, SisBaHiA [20], for hydrosedimentological aspects in the adjacent coastal region and the Soil & Water Assessment Tool, SWAT [21], for hydrological modeling of the river basins of interest.

SisBaHiA, developed at COPPE|UFRJ since 1987, has been expanded and improved over the years, being adopted in several studies and projects related to the modeling of water bodies. SWAT, on the other hand, under continued development by the Agricultural Research Service and Texas A&M University since 1996, is widely used to calculate flows and loads of water, nutrients, and sediments, based on specific information about water, soil properties, topography, vegetation, and soil management practices in the watershed. The combination of these models allowed for detailed simulations of river flows and suspended sediment concentrations, providing a solid basis for environmental assessments and decision-making.

2.1 Hydrological modeling system: SWAT

Considering the lack of data on TSH concentrations in the Paraíba do Sul and Itabapoana rivers, it was proposed to estimate these concentrations using the public domain system SWAT—Soil & Water Assessment Tool.

SWAT is a time-evolving model and usually works on a daily time frame. One of its advantages is the possibility of making simulations for short or long time series. It considers the watershed divided into sub-basins based on relief, soils, and land use and thereby preserves the spatially distributed parameters of the entire basin and the homogeneous characteristics within the basin. The input data for each sub-basin can be grouped or organized according to the following categories: climatic, Hydrologic Response Units (HRUs), ponds/wetlands, groundwater, main water course, and sub-basin drainage. To generate the HRUs in each sub-basin, a sensitivity level is adopted, eliminating land use classes with an area smaller than the established value [22].

The implementation of SWAT requires several input parameters for a coherent representation of the physical characteristics of a watershed. In the specific case of the present study, a hydrological model was developed with two outlets, corresponding to the hydrographic basins of the Paraíba do Sul and Itabapoana rivers (**Figure 1**).

In the development of hydrological modeling in SWAT, the parameters were divided into Digital Elevation Model, land use and occupation data, and pedological and meteorological data for each watershed, as detailed below.

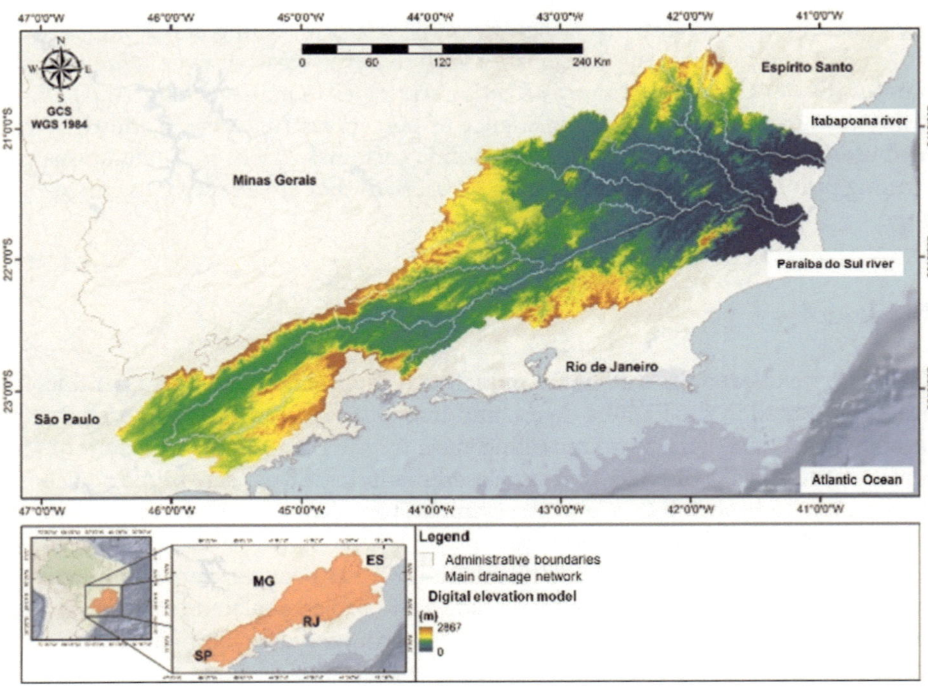

Figure 1.
Digital elevation model with the identification of the modeled hydrographic basins and their respective main drainage.

2.1.1 Digital elevation model

To simulate the physical and chemical processes of a river basin, we start with the MDE. It is necessary for the mathematical representation of the altimetric surface to be studied, through the coordinates X, Y, and Z. The MDE was a mosaic that originated from the composition of 21 images of the ASTER Global Digital Elevation Model V003 satellite [23] with a resolution of 30 m. To fulfill the objective of the present study, the hydrographic basins modeled were the basins of the rivers: Paraíba do Sul and Itabapoana, as shown in **Figure 1**.

2.1.2 Land use and occupation map

In the process of occupying a territory, several natural and cultural factors are preponderant. And in this way, the configuration of land use and occupation directly influences the ecological systems of the region, producing effects on its landscape. Knowledge of land use and occupation is fundamental for space quality management, as well as essential for understanding the erosion processes and susceptibility to erosion of an area. For the SWAT modeling, a mosaic composed of 04 Vector Thematic Bases of Land Cover and Land Use of the states of Espírito Santos, Minas Gerais, Rio de Janeiro, and São Paulo of the Brazilian Institute of Geography and Statistics—IBGE for the year 2018 was used at a scale of 1:1,000,000, cf. **Figure 2**. It should be noted that the data were obtained from the website of the National Spatial Data Infrastructure (INDE) [24].

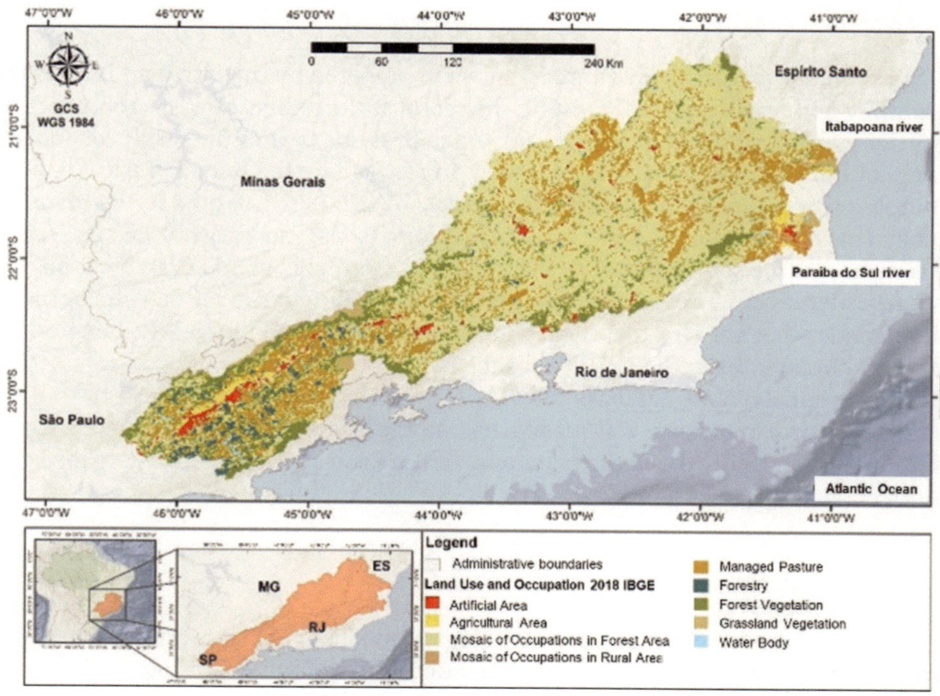

Figure 2.
Land use and occupation map used in hydrological modeling from the vector thematic bases of land cover and land use of the states of Espírito Santos, Minas Gerais, São Paulo, and Rio de Janeiro of the IBGE for the year 2018.

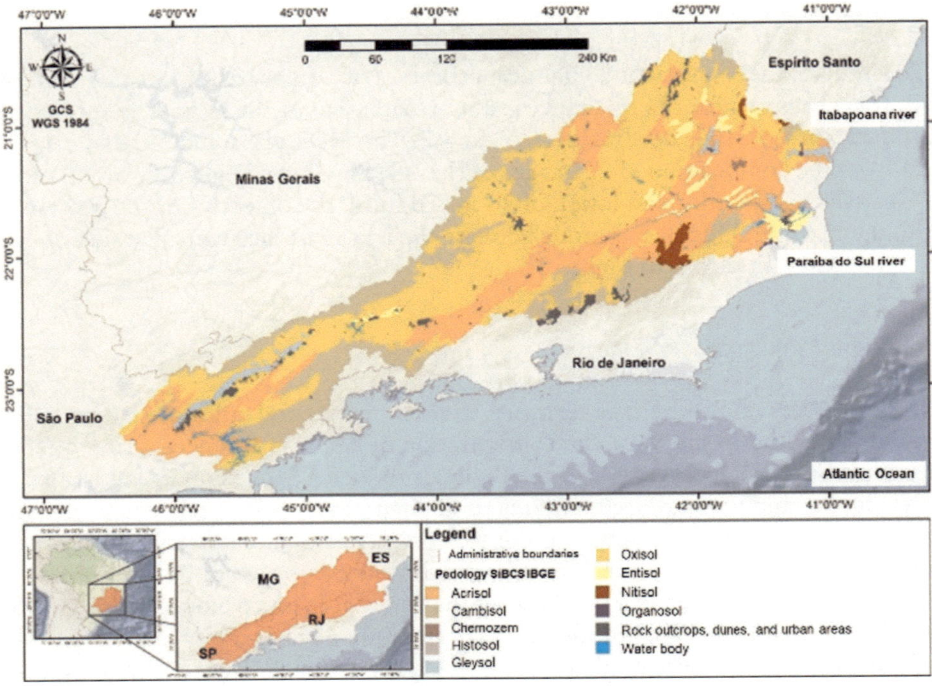

Figure 3.
Pedological map used in hydrological modeling from the publication macrocharacterization of natural resources of Brazil, prepared by IBGE, with classification of soils at the order level.

2.1.3 Pedological map

The pedological data used for modeling were prepared by the Brazilian Institute of Geography and Statistics – IBGE within the scope of the publication Macrocharacterization of Natural Resources of Brazil, which presents part of the results obtained throughout the Natural Resources Mapping project, where the Brazilian natural resources were mapped and described, at a scale of 1:250,000, related to the themes geology (rocks), geomorphology (relief), pedology (soils), and vegetation. The soil types represent the soil classes of Brazil, based on the Brazilian Soil Classification System (order level). These classes are grouped according to similar morphological characteristics that are repeated throughout the country; see **Figure 3**. Pedological data were obtained from the INDE website [24].

As the original SWAT database refers to the standards of North American soil parameters, correspondence was made between the classes of the Brazilian Soil Classification System (SiBCS) and the taxonomy of the Food and Agriculture Organization (FAO), cf. supervised by Embrapa Soils [25].

2.1.4 Weather data

The meteorological data used for hydrological modeling were: precipitation, minimum and maximum temperatures, relative humidity, and wind and solar radiation. The locations of the stations are distributed in the modeling domain as shown in **Figure 4**. These data were obtained through the POWER Project through the electronic address POWER | Data Access Viewer [26]. This project provides solar and

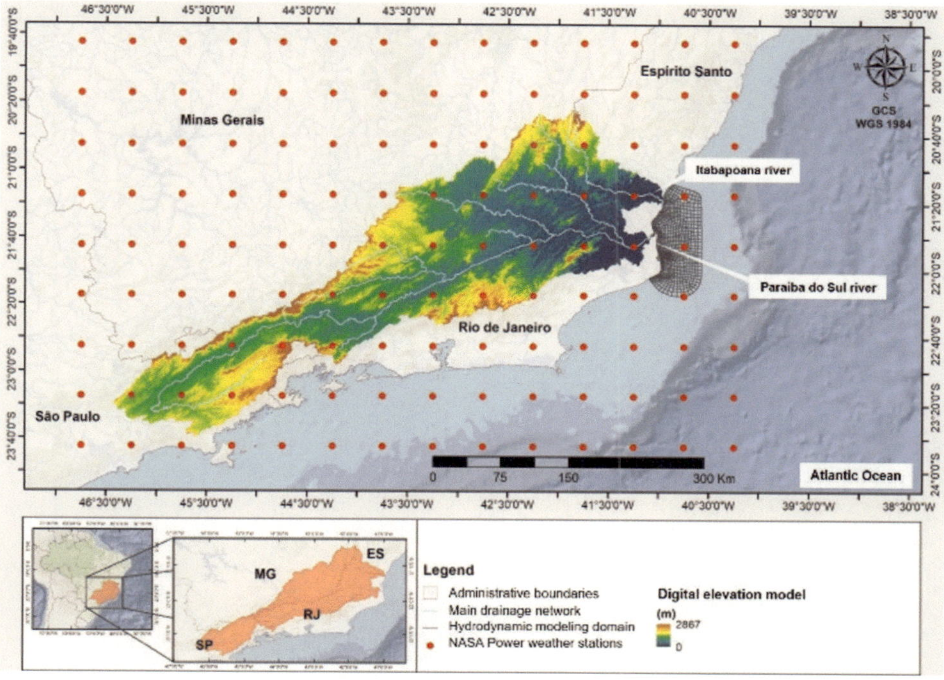

Figure 4.
Weather data stations extracted from POWER | Data Access Viewer from NASA Prediction of Worldwide Energy Resources.

weather datasets sourced from NASA research to support renewable energy, building energy efficiency, and agricultural needs. According to [27], the project's products are satellite-based, and models have been shown to be accurate enough to provide reliable solar and meteorological resource data in regions where surface measurements are scarce or non-existent. Documentation of the project and its methodology can be found on the NASA POWER | Docs [28].

2.2 Hydrosedimentological modeling system: SisBaHiA

The models used in the hydrosedimentological analyses of the adjacent coastal zone discussed in this study are part of SisBaHiA—Base System of Environmental Hydrodynamics. Technical details about the methodology adopted in the SisBaHiA models as well as the mathematical and numerical formulation of the system models and the imposition of boundary conditions can be seen in [29].

The use and availability of the models developed in SisBaHiA are obtained free of charge through SisBaHiA [20], where you will find instructions on how to install the system. The overall strategy of the system is based on a hierarchical database system, in order to provide reliability and ease of manipulation of data and results to users.

In the development of hydrosedimentological modeling in SisBaHiA, the parameters were divided into Digital terrain modeling, Fluvial and sedimentological constraints, and Meteorological and oceanographic constraints. The quantification of sedimentation was done through hydrosedimentological computational modeling with six coupled models running simultaneously, namely: Hydrodynamic Model, Salinity and Temperature Transport Model, Non-Cohesive Sediment Model, Cohesive

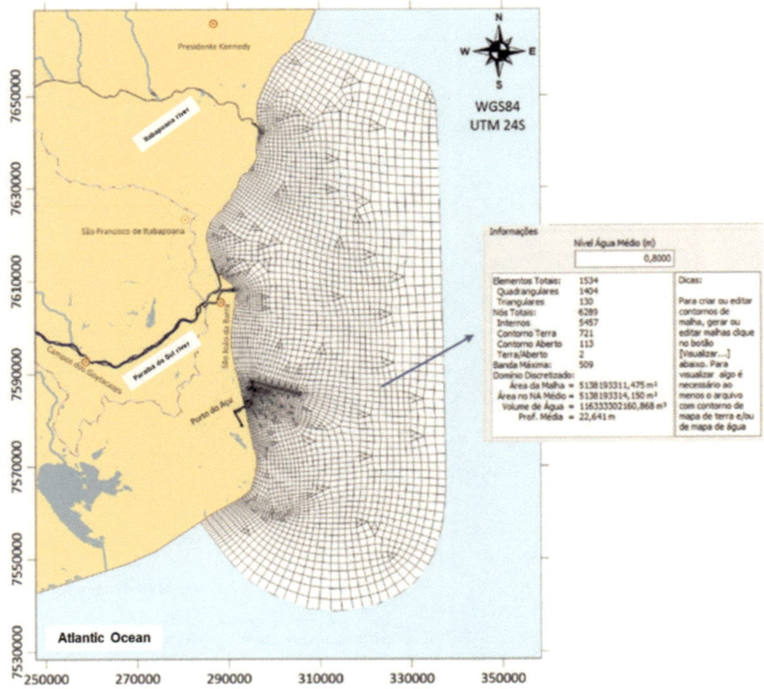

Figure 5.
Discretization of the modeling domain into triangular and quadrangular subparametric finite elements. Triangular elements are quadratic, and quadrangular elements are biquadratic.

Sediment Model, Wave Model and Generation with local winds, and Wave Propagation Models generated outside the local domain.

2.2.1 Digital terrain modeling

In the development of maps to delineate the modeling domain, we move on to the discretization of this domain using a finite element mesh; a detailed approach is illustrated in **Figure 5**. This step is complemented by the inclusion of informative layers, with bathymetry data presented in **Figure 6**, and the definition of the amplitude of the equivalent roughness of the bottom, varying according to the material present in the bed.

2.2.2 Fluvial and sedimentological constraints

The analysis of the liquid and solid flows from the Paraíba do Sul and Itabapoana rivers, covering both fine suspended sediments and more granular forms, reveals important nuances about the sedimentary processes of the region. Using the SWAT model to calculate both the liquid flows, illustrated in **Figure 7**, and the concentrations of suspended sediments (**Figure 8**), we focused on components such as clays and fine silts. These fine sediments, often categorized as Total Suspended Sediments (TSS), are essential to understanding sedimentary transport in the water column, highlighting the complexity and dynamics of the river systems under study.

In the context of sedimentology of fluvial and estuarine environments, modeling the solid flows of sediments, such as coarse silt, sands, and even gravel, is important.

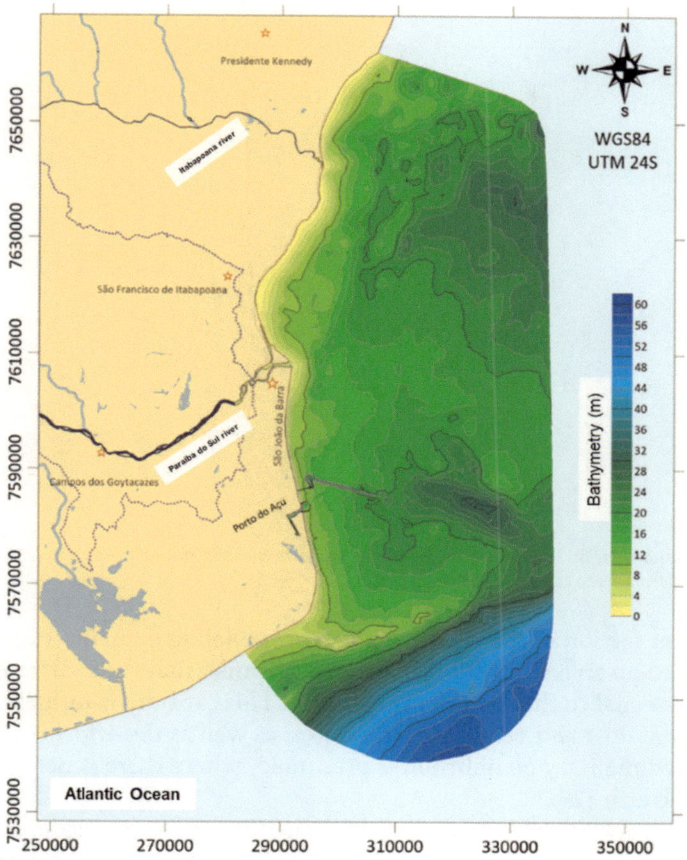

Figure 6.
*Bathymetry map resulting from digital modeling of the domain of interest in a finite element mesh, as shown in the map in **Figure** 5.*

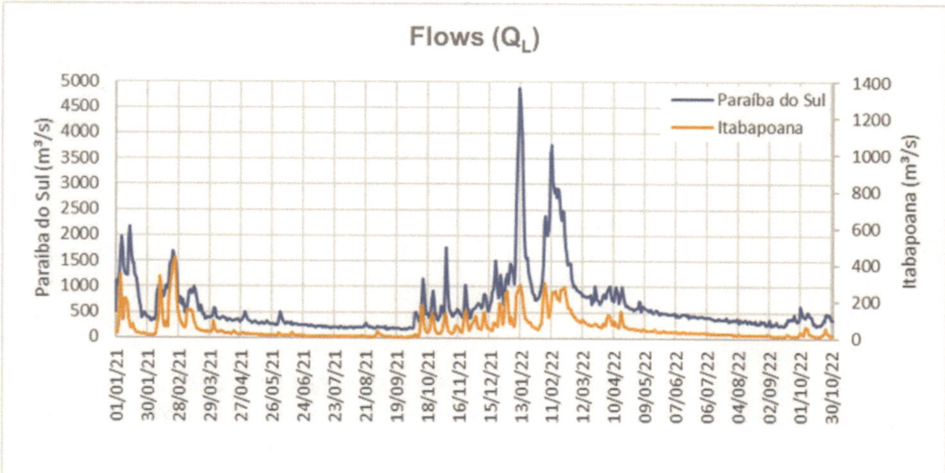

Figure 7.
Hydrographs of simulated net flows for the hydrographic basins of the Paraíba do Sul and Itabapoana rivers by the SWAT hydrological model from 01/01/2021 to 10/31/2022.

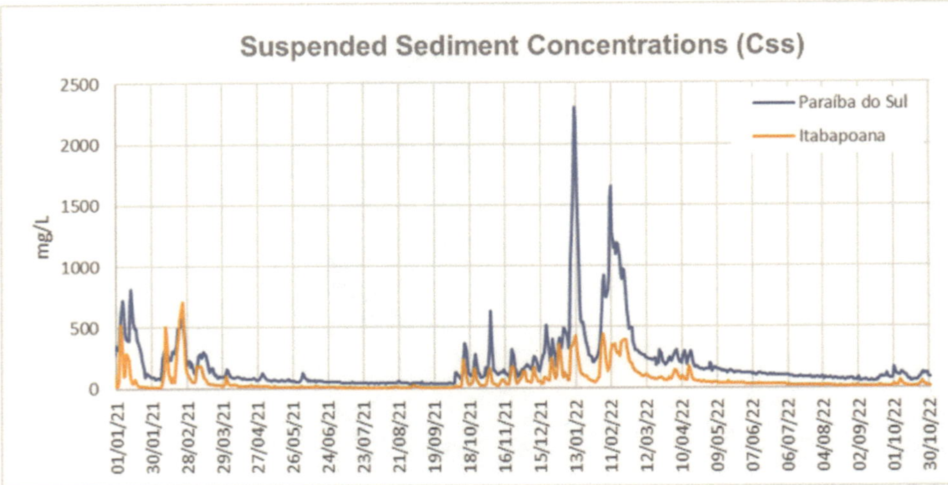

Figure 8.
Hydrographs of simulated TSS concentrations for the Paraíba do Sul and Itabapoana river basins by the SWAT hydrological model from 01/01/2021 to 10/31/2022.

In the upstream section of the rivers, within the modeling domain, these flows are calculated based on an equilibrium condition, assuming that the sediment load carried by the rivers is equal to their transport capacity. This capacity is influenced by variables such as the flow and velocity of the water, as well as the effects of the tide. Therefore, a sedimentary equilibrium is presumed, where there is no deficiency of sediments upstream [29].

For the non-cohesive sediments, a spatial distribution was considered according to sedimentary typologies described by [30]. Thus, the sedimentological characterization of the region was based on previous studies that mapped the distribution of sediments, providing a critical reference for the understanding of sedimentary processes.

2.2.3 Meteorological and oceanographic constraints

In this section dedicated to meteorological and oceanographic conditions, the study details the use of data from five stations of the FES, Finite Element Solution [31], and six stations of the HYCOM, Hybrid Coordinate Ocean Model [32], to generate data on astronomical and meteorological tides.

The data were applied to all 113 maritime boundary nodes defined as “Open Contour.” This procedure allowed a refined understanding of the influences of tides on the studied system, essential for hydrodynamic and sediment transport analyses, highlighting the interconnection between oceanographic and meteorological phenomena in marine environmental modeling. The locations of the stations used are illustrated in **Figure 9**.

Regarding astronomical tides, we highlight the methodology used to distribute their effects along approximately 200 km of maritime border. This process was carefully carried out by weighting values from five stations in the FES2014, thus ensuring an accurate and detailed representation of the rise in astronomical tide levels (**Figure 10**).

In the modeling process, described sea level variations attributable to non-astronomical factors, commonly referred to as meteorological tides, were mapped

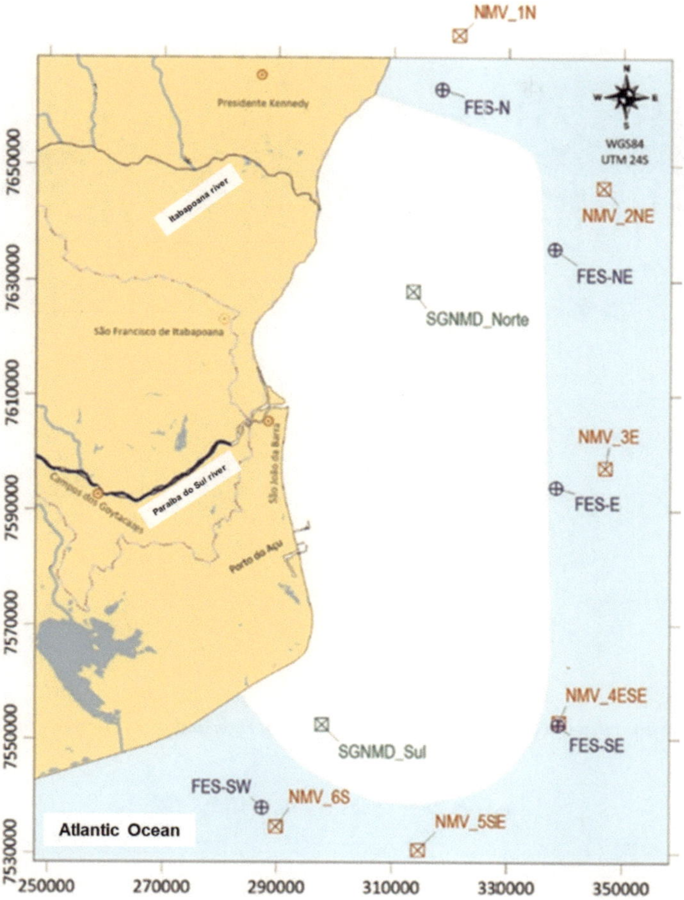

Figure 9.
Positioning of the five stations of harmonic constants of astronomical tides of the FES and the six stations of the HYCOM with values of NMV—variable mean level and of the SGNMD—differential mean level generator series stations for induction of drift currents.

along approximately 200 km of maritime boundary. This distribution was performed using data from six Variable Mean Level (NMV) stations, obtained from HYCOM (**Figure 11**). These data were integrated with astronomical tidal data, strengthening the model's accuracy in predicting sea level variations.

The induction of drift currents due to meteorological effects was performed by incorporating HYCOM data from two SGNMD stations located on the continental shelf (**Figure 9**). This process involved the weighted assimilation of these data, introducing them into the Hydrodynamic Model of SisBaHiA as boundary conditions of Differential Mean Level. The coastal drift currents resulting from this methodology were characterized by distinct directions, positive to the east and negative to the west, reflecting the coastal dynamics influenced by meteorological variations (**Figure 12**).

Salinity and temperature data were collected using the HYCOM model for the study period. These data, which are important for the analysis of oceanographic conditions, were extracted at the specific points of Mean Variable Level (NMV). The graphical representation of these data, as illustrated in **Figure 13**, reflects the mean values in the water column, providing a clear view of thermohaline variations over time.

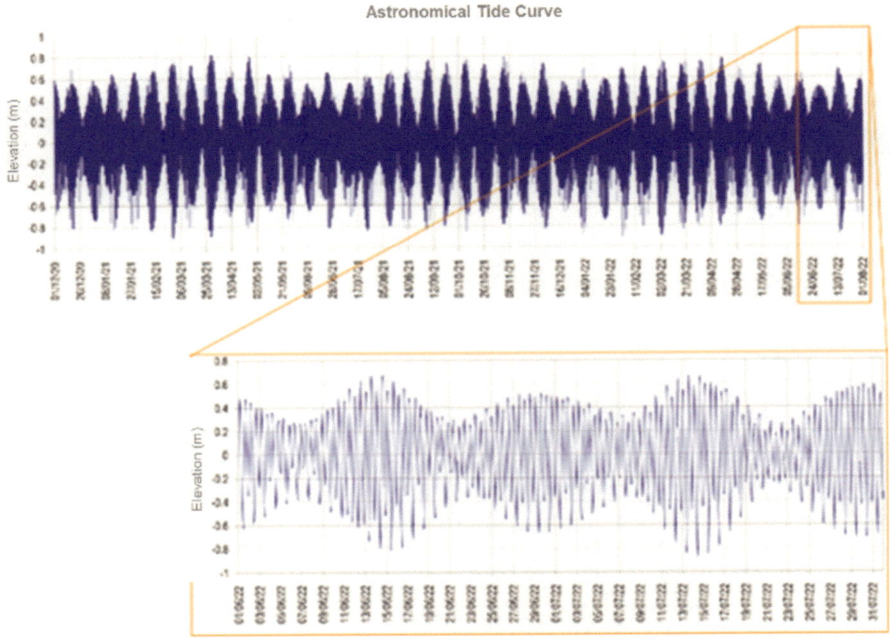

Figure 10.
Elevation of the astronomical levels imposed on the model at the open boundary of the modeling domain.

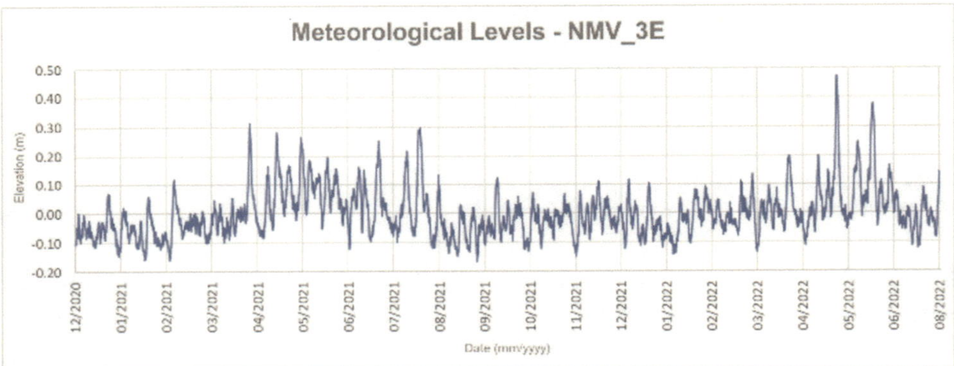

Figure 11.
Meteorological levels imposed on the model at the open boundary of the modeling domain. The above series refers to station 'NMV_3E' only. Similar graphs can be produced for all variable medium level stations.

To incorporate the dynamic effects of winds into the oceanographic study, we used ERA5 reanalysis data [33], detailing the intensity and direction of the winds as shown in **Figure 14**. This procedure significantly enriches the modeling, allowing a more accurate simulation of meteorological conditions and their influences on maritime dynamics. The data, sourced from the ECMWF (European Centre for Medium-Range Weather Forecasts) through the ERA-Interim and ERA5 reanalysis, cover the years 2021 and 2022, providing a solid basis for the analysis of wind impacts on oceanographic conditions.

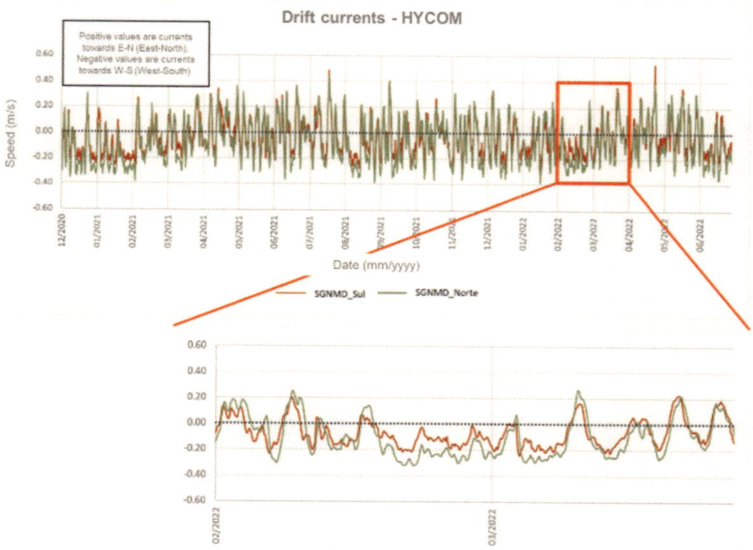

Figure 12.
Coastal drift currents obtained from HYCOM and inserted as differential mean level boundary conditions in the SisBaHiA hydrodynamic model. Positive values indicate currents to the east and negative values to the west. See position of SGNMD inductors in ***Figure 9****.*

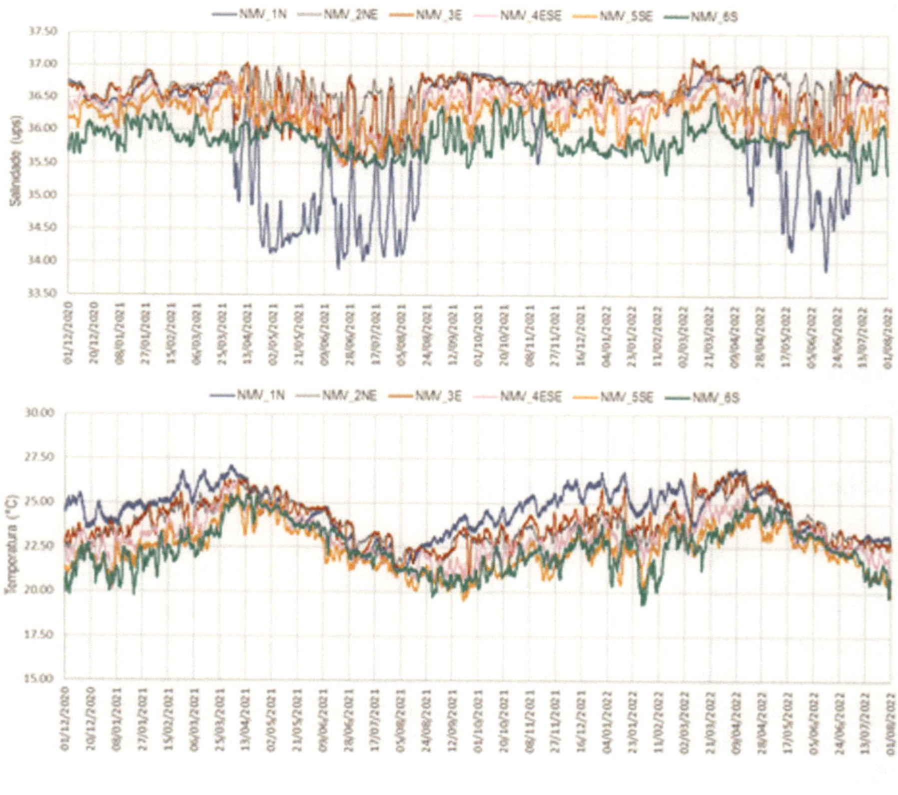

Figure 13.
Salinity and temperature series at the stations illustrated in ***Figure 9****.*

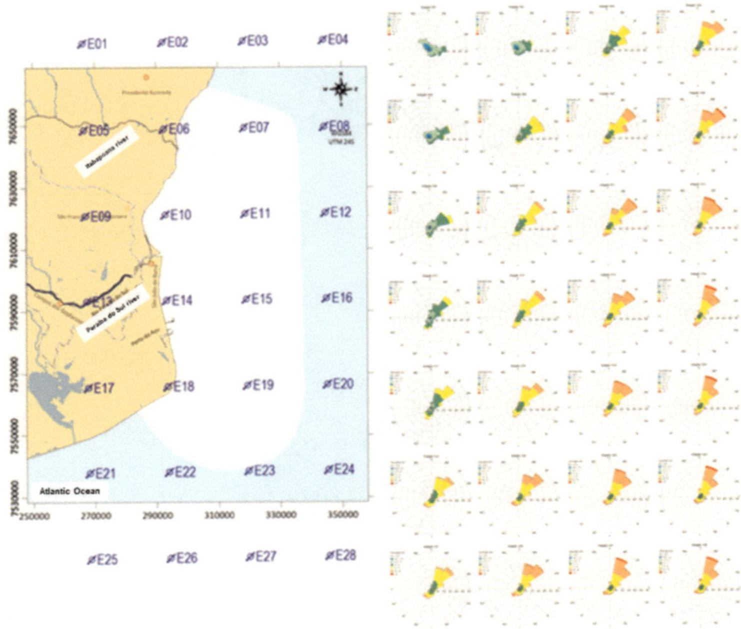

Figure 14.
Map indicating wind positions and compass rose patterns for each season, based on wind data sourced from ECMWF reanalysis—ERA-Interim and ERA5, covering the years 2021 and 2022.

The action of winds in the modeling domain acts on the induction of drift currents and the generation of waves. The waves, in turn, induce radiation voltages that act on the current field in the shallower coastal zone, especially in the surf zone. Wave data were extracted from the ERA5 reanalysis model, run by ECMWF [33].

The data obtained constitute a time series of significant wave height, peak period, and average wave direction, with information every 1 hour, so the model minimally considers the interference of tidal level oscillation in wave propagation.

Analyzing the wave data, it can be seen that the waves with the highest probability of occurrence are from the southeast and south quadrants, between 110° and 200°, with a peak period predominantly between 8 and 12 s and a significant height between 1.5 and 2.0 m.

2.3 Calibration and validation of computational models

The calibration and validation of computational models represent fundamental steps in the development of reliable and accurate simulations of hydrosedimentological processes. In them, we ensure that the models are able to reproduce natural processes with fidelity, considering the complex interactions between the hydrological and sedimentary components of the systems of interest [34]. Calibration involves adjusting model parameters until the outputs match as closely as possible to the observed data, while validation tests the model's ability to predict conditions different from those used in calibration [35].

Below we present the methodologies applied in the calibration and validation of the computational models used in this study, SWAT and SisBaHiA.

2.3.1 SWAT

Calibration and validation were performed automatically using the Calibration and Uncertainty Procedures (SWAT-CUP) program [36], specifically the Sequential Uncertainty Fitting (SUFI2) algorithm. Further details on the calibration and validation procedures are detailed in the user manual of the aforementioned program—SWAT-CUP: SWAT Calibration and uncertainty programs—A user manual [37]. In **Figure 15**, it is possible to observe the connection of the hydrodynamic modeling domain, SisBaHiA, with the hydrological modeling domain, SWAT, for each hydrographic basin of interest—Paraíba do Sul and Itabapoana.

In SWAT, the maximum simulation period is determined by the comprehensiveness of the climate data. In the study in question, for the liquid flows, the most consistent data to perform the calibration and validation were those of the stations: 57930000, Santa Cruz in Mimoso do Sul on the Itabapoana River and 58974000, Campos—Municipal Bridge in Campo dos Goytacazes on the Paraíba do Sul River, cf. illustrated in **Figure 15**. The calibration period was from 01/01/2012 to 12/31/2015 and the validation period from 01/01/2016 to 12/31/2018.

For TSS concentrations, the analyzed data available from ANA at the calibration stations showed inconsistent results, as described by [38]. In order to compare the modeled TSS concentrations, the station used as a reference was the 58385100 code HPP, Simplício Montante HPP of the National Water Agency (ANA) located in Três

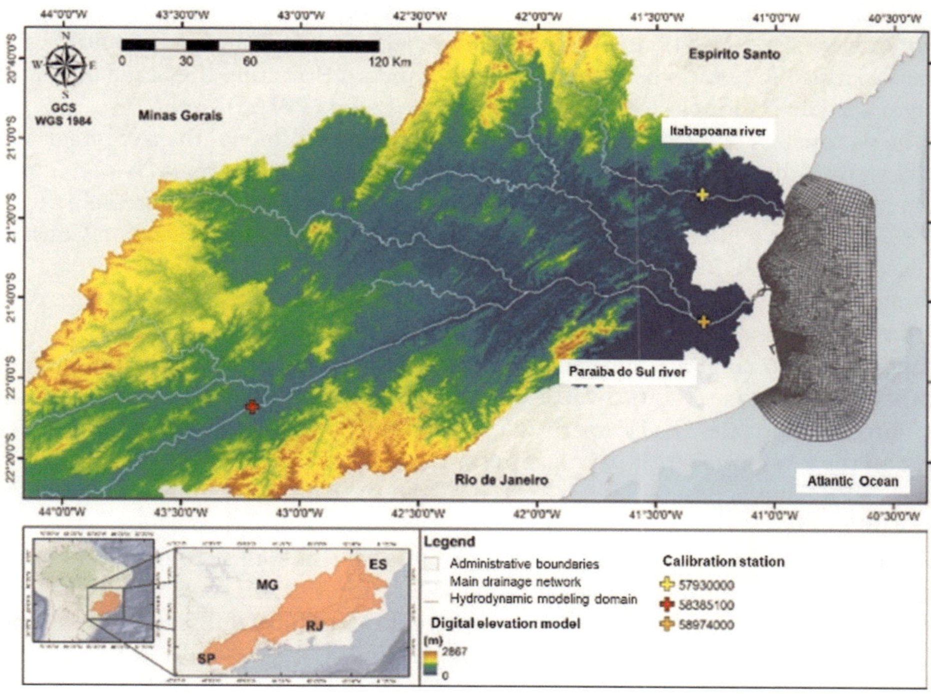

Figure 15.
Connection of the hydrodynamic modeling domain, SisBaHiA, with the hydrological modeling domain, SWAT, for each watershed of interest—Paraíba do Sul and Itabapoana. And location of the calibration stations for liquid flow—57930000, Santa Cruz in Mimoso do Sul on the Itabapoana River; 58974000, Campos—Municipal Bridge in Campo dos Goytacazes on the Paraíba do Sul River and for comparison of the concentration of OSH—58385100, Simplício Montante HPP in Três Rios on the Paraíba do Sul River, National Water Agency (ANA).

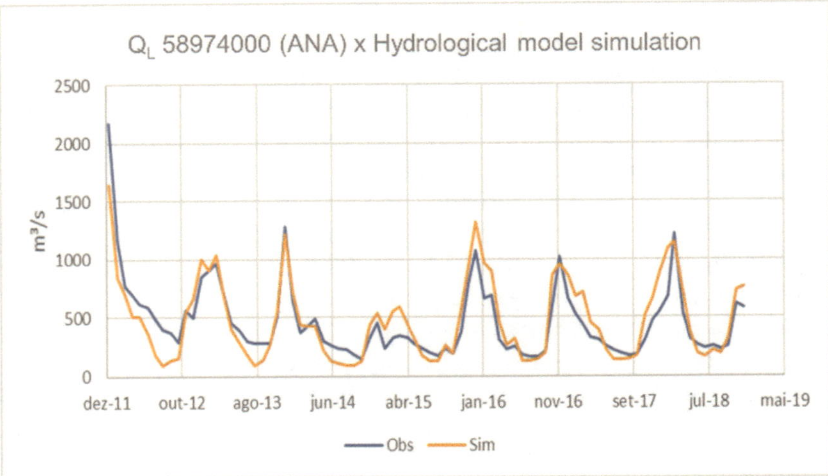

Figure 16.
Calibration: Illustration of the hydrograph of the liquid flow of the calibration station 58974000, Campos—Municipal Bridge in Campo dos Goytacazes—National Water Agency (ANA) and calibrated hydrograph obtained by the hydrological model (SWAT).

Rios in the Paraíba do Sul River Basin. The suspended sediment transport curve used was $Q_{ss} = 0,0263 \times Q^{2,116}$ [38].

In order to evaluate the efficiency of the simulations, the following performance statistics were used: Nash-Sutcliffe Efficiency Coefficient (COE), Coefficient of Determination (R^2), Ratio between the Mean Standard Error and the Standard Deviation of the Observations (RRR), and Percentage Trend (PBIAS) [39].

The parameters adjusted in the calibration of the liquid flow of the hydrological model were: Value of the Number Curve for the humidity condition II (NC2); Limit depth of the water level above which the base flow is released by the channel (GWQMIN); Time interval for aquifer recharge (GWDelay); and Recession Constant for Base Flow (AlphaBF).

From **Figures 16–21**, it is possible to observe that the model simulated, in general, very well the values of net flows for the BH of the Paraíba do Sul River. For the Itabapoana river basin, the simulated period can be considered satisfactory; however, the statistics present worse classified values than in the previously mentioned watershed.

Figure 22 illustrates the comparison of the daily TSS concentration calculated based on the key curve of [38] applied to the net flow of the 58385100 comparison station, HPP Simplício Montante in Três Rios, and the daily flow simulated by SWAT after calibration.

Under the conditions exposed, the results presented, as well as previous studies described, we consider that the result of the TSS concentrations obtained by the hydrological model and its variability are coherent. Therefore, utilizing their findings as input for the SisBaHiA hydrosedimentological model.

2.3.2 SisBaHiA

The critical analysis of model calibration and validation procedures, as outlined by [29], highlights the importance of the accuracy of the input data for the reliability of

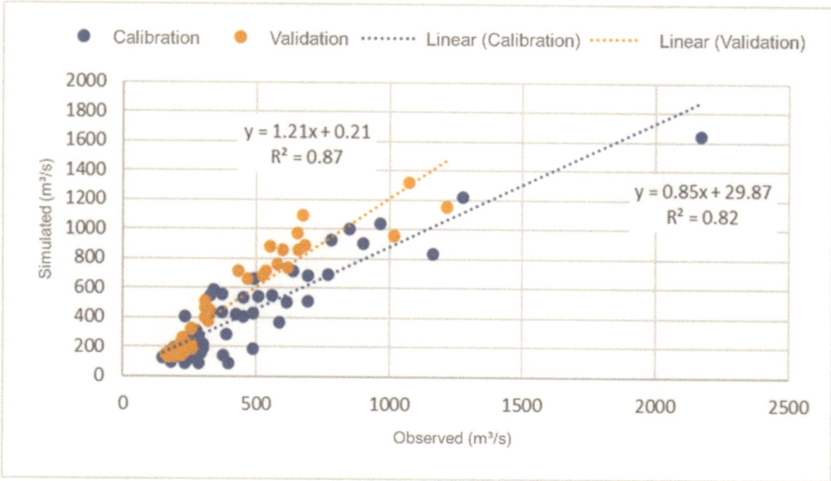

Figure 17.
Monthly net flow (m³/s) simulated and observed for the calibration and validation periods, Paraíba do Sul River.

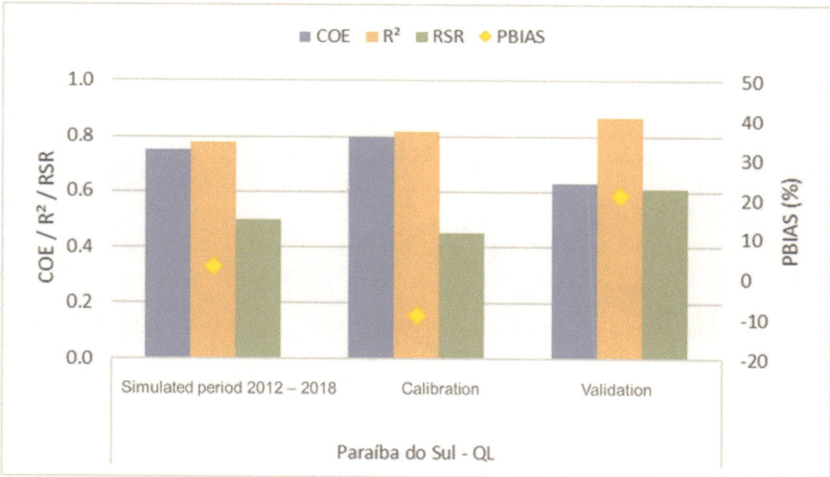

Figure 18.
Performance statistics of the calibrated and validated model, Paraíba do Sul River.

the modeling results. SisBaHiA models demonstrate high coherence with field data in multiple aspects, including shoreline geometry, bathymetry, water levels, and current velocity, with significant improvements after calibration. Accuracy in modeling sedimentological processes and wave propagation depends on correct formulation and accurate particle size data. The conditions to achieve high coherence involve the accuracy of the geometry of the water body, flow forcings, intensities of contaminant sources, and sedimentological variables.

The calibration of the environmental hydrodynamics models in SisBaHiA was a process that followed the following order:

Level 1. Geometric calibration

> Comparative maps of Level 1 calibration are omitted, related to geometry, bathymetry, and roughness amplitudes, since they result from direct data

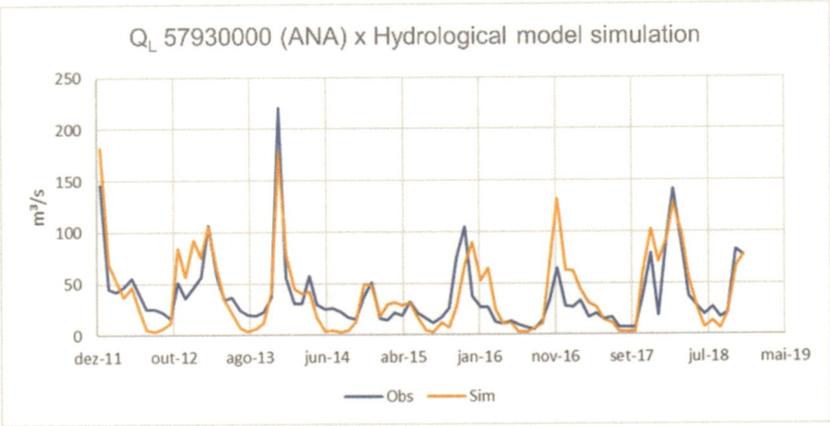

Figure 19.
Hydrograph of the liquid flow of calibration station 57930000, Santa Cruz in Mimoso do Sul—National Water Agency (ANA) and calibrated hydrograph obtained by the hydrological model (SWAT).

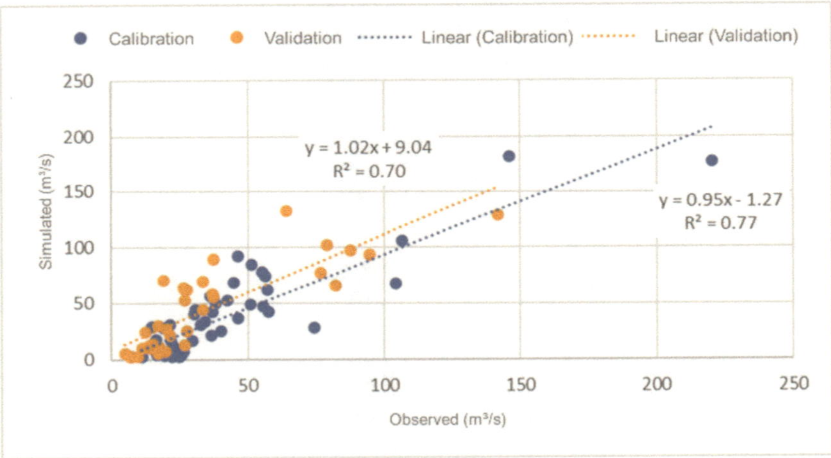

Figure 20.
Monthly net flow (m³/s) simulated and observed for the calibration and validation periods, Itabapoana River.

interpolation and are duly verified. As exposed, the domain of interest is geometrically adequate, as well as the discretization of the mesh is sufficiently refined to capture effects of interest.

Level 2. Hydrodynamic and sedimentological calibration

The first step in the level verification process focuses on defining the boundary conditions at the open border. The record made at station T1—Access Bridge [40] was adopted as the basis for generating records of tide levels along the open border nodes. The station was used to check tide levels from 21/12/2020 to 31/10/2022.

Figure 23 shows that, despite the small discrepancies between the measured Average Daily Levels (NMDia) and those calculated at the T1—Access Bridge station, they do not significantly influence the astronomical tidal amplitudes. These variations

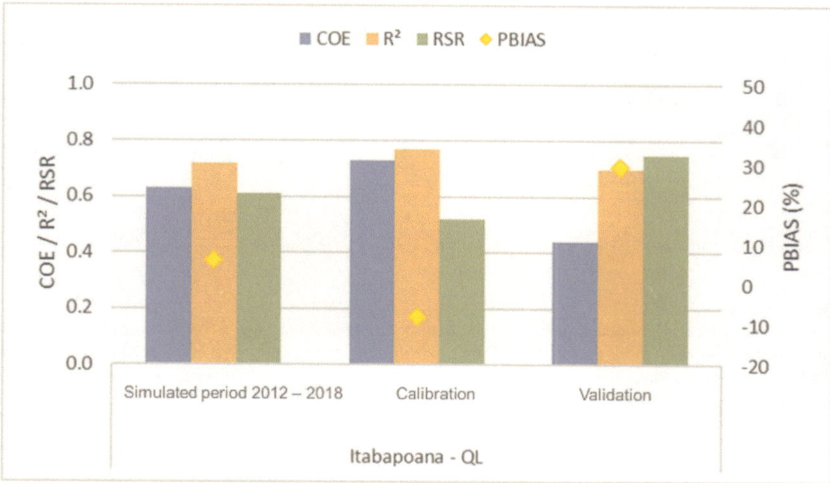

Figure 21.
Performance statistics of the calibrated and validated model, Itabapoana River.

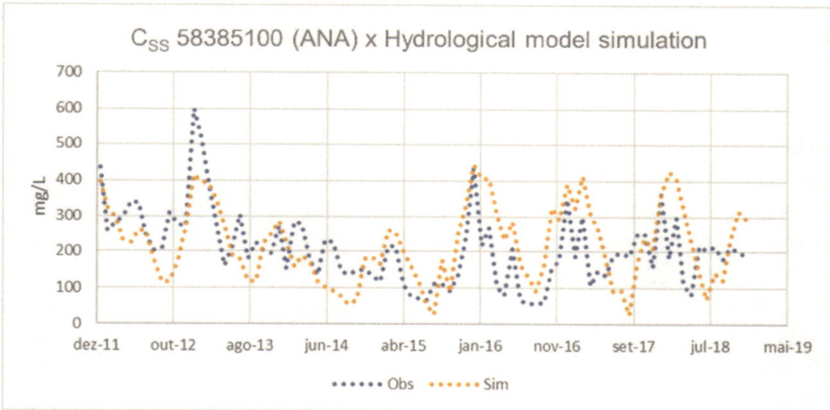

Figure 22.
Hydrograph of the TSS concentration calculated based on the key curve of the preliminary diagnosis of the hydrosedimentological conditions of the Paraíba do Sul River and its main tributaries (1998) applied to the net flow of the 58385100 comparison station, HPP Simplício Montante em Três Rios—Agência Nacional de Águas (ANA) and calibrated hydrograph obtained by the hydrological model (SWAT).

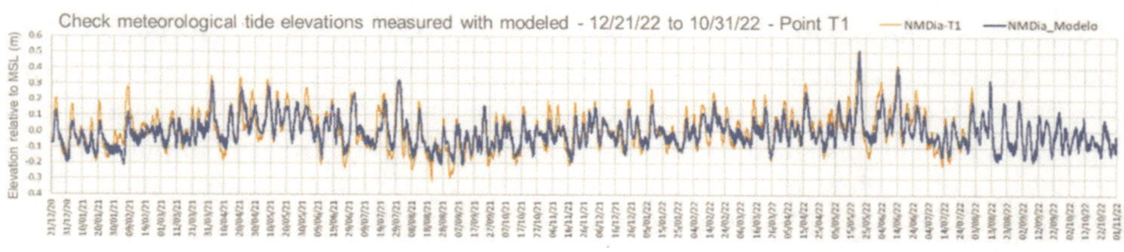

Figure 23.
Verification of meteorological tides that modify the average daily level (NMDia) obtained from measurements at the T1 station—Access bridge and values calculated by the model, between 12/21/20 and 10/31/22. The measured data available was up to 07/31/2022. Inequalities in NMDia cause the small differences on some days between measured and modeled tidal elevations shown in ***Figure 24****.*

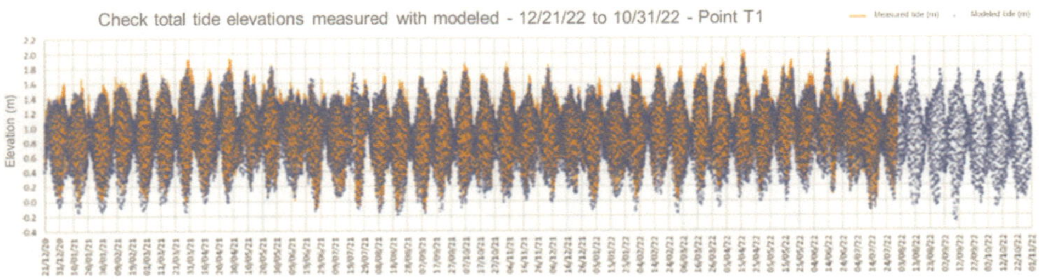

Figure 24.
Verification of measured and modeled tide levels at station T1—Access bridge, between 12/21/20 and 10/31/22. The measured data available were up to 07/31/2022. Analyzing the graphs, it is evident that the modeling results are visually very close to the measured values. The slight differences are due to inaccuracies in meteorological tidal variations obtained from HYCOM results.

are attributed to the Variable Mean Level (NMV) conditions at the open frontier, derived from the HYCOM results. In addition, **Figure 24** reveals that the tidal amplitudes, both measured and modeled, have great similarity, although punctual variations in the maximum or minimum elevations may occur due to meteorological tides, affecting the NMDia.

3. Results

In this segment of the book, we devote attention to the analysis of the results from the modeling and analysis of data that illustrate the hydrological and sedimentological processes acting in the hydrographic basins of the Paraíba do Sul River, Itabapoana, and in the adjacent coastal region, with special focus on the terminals of the Port of Açu.

The Port of Açu is a port and industrial complex, operating since 2014 in deep waters. It plays an important role in operations related to oil and gas, mining, logistics, and the energy sector, standing out for its ability to accommodate large ships thanks to its considerable depth. In the environmental context, the maintenance of its navigability and the expansion of its activities imply dredging procedures, which are essential for the control of siltation and the guarantee of adequate depth for port operations.

The results presented show the dynamics of water flows and suspended sediments in the period from 01/01/2021 to 10/31/2022, highlighting how watersheds respond to extreme events and the influence of meteorological phenomena, especially the South Atlantic Convergence Zone (SACZ). This analysis allows us to understand the silting and erosion processes that impact the terminals of the Port of Açu. Throughout this chapter, we have emphasized the observed impacts, highlighting the importance of using integrated models in the efficient management of water resources and in the mitigation of adverse environmental impacts.

The results obtained in the SWAT hydrological modeling, both for the net river flows and the TSS concentrations, were used as input data for the hydrosedimentological modeling carried out by SisBaHiA for the period from 01/01/2021 to 10/31/2022. These net flows and TSS concentrations for the rivers of interest, Paraíba do Sul and Itabapoana, are illustrated cf. **Figures** 7 and **8**.

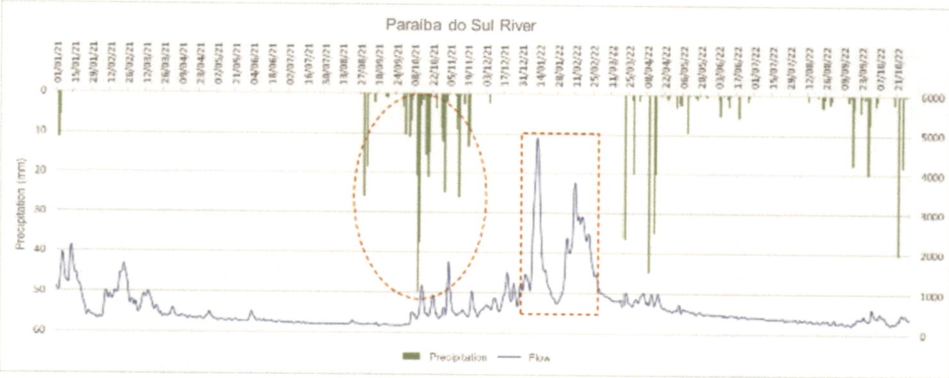

Figure 25.
Precipitation (mm) and flow (m^3/s) records recorded at the 58974000 Campos—Ponte Municipal station, from 01/01/2021 to 10/31/2022. ○ *Significant precipitation records (mm) and* □ *significant flow records (m^3/s) were highlighted.*

We will proceed with the identification of extreme events and hydrological disasters that occurred between 01/01/2021 and 10/31/2022, an essential step to understand the responses of the river basins and the Port of Açu to such events. Hydrological modeling plays a significant role in the analysis of these responses, directly influencing the management and sustainability of the port. This discussion aims to elucidate how hydrological processes affect both sedimentary dynamics and port operations, emphasizing the importance of integrating hydrological modeling and environmental management practices.

In order to understand the dynamics of extreme precipitation events and their evolution to hydrological disasters in the northern region of the state of Rio de Janeiro, it is essential to analyze the relationship between the recorded precipitation and the variations in the levels of the Paraíba do Sul and Itabapoana rivers. The choice of the station 58974000 Campos—Ponte Municipal as the focal point allows us to adequately illustrate the rainfall and hydrological behavior of the Paraíba do Sul River during the period studied (**Figure 25**).

The data collected indicate that the extreme rainfall event that occurred at the end of 2021 was the most significant in the analyzed interval. However, it is interesting to note that the maximum flows, which culminated in a hydrological disaster in the region, were recorded in the months of January and February 2022, demonstrating an apparent disconnect with the immediate local precipitation.

This observation suggests that the hydrological disaster must be associated with a broader event, involving the Paraíba do Sul River basin as a whole. It is important to highlight that this basin extends through the states of Minas Gerais, Rio de Janeiro, and São Paulo, covering a considerable and complex area. Therefore, the analysis of extreme events and hydrological disasters requires an integrated view that considers the contributions of the entire watershed, reflecting on the interactions between the various climatic and geographic factors that influence the behavior of rivers in this region.

The analysis of meteorological information provided by the National Institute of Meteorology [41] through the Technical Note on the extreme events of February 2022 in Brazil provides a solid basis for understanding the hydrological disasters in the northern region of the ERJ and their connection with the Paraíba do Sul River basin. The intense rains, with episodes concentrated in the states of Rio de Janeiro and Minas

Gerais, both integral parts of the watershed, highlight the regional interconnection in the observed impacts.

The events described resulted in severe consequences, including loss of life, flooding, and landslides. The role of the South Atlantic Convergence Zone (SACZ) was unique, with episodes that provided rainfall accumulations above the historical average, in line with the significant flows observed. Specifically, the exceptional accumulated rainfall in Itaperuna, with 436.7 mm with a significant deviation of 325.9 mm from the normal February average in the state, which is 110.8 mm, illustrates the magnitude of the events and their direct influence on river dynamics and subsequent disasters.

This set of evidence reinforces the conclusion that the observed hydrological disasters are linked to comprehensive meteorological phenomena and their influence on the Paraíba do Sul River basin, evidencing the need for an integrated management of water resources that considers the complex meteorological and hydrological interactions.

The analysis of the Itabapoana River basin, using the station 57836000 PCH Pedra do Garrafão Downstream, reveals a hydrological response that is directly aligned with the local rainfall accumulations, in contrast to the observations made for the Paraíba do Sul River basin. This behavior indicates a more immediate correlation between local rainfall and hydrological variations in the Itabapoana basin, as illustrated by **Figure 26**. In addition, the influence of the SACZ on the accumulated rainfall in February 2022 was significant, evidencing the impact of this meteorological phenomenon in this region as well.

In February 2021, the Itabapoana basin experienced a period of intense rainfall, corroborated by the Joint Alert for Heavy Rainfall in the Southeast Region [42] issued by INMET for the Southeast Region of Brazil, which was under the influence of the SACZ. The persistent, stationary nature of the SACZ, coupled with the associated rainfall pattern, resulted in heavy rainfall in a short period of time, accumulating high volumes of water over the days.

These extreme precipitation events were efficiently captured by the hydrological modeling performed, as shown in **Figures 7** and **8**. This observation underlines the relevance of employing comprehensive watershed modeling, which considers not only local variables but also the impacts of broader meteorological phenomena such as the SACZ. Such an approach is indispensable for a holistic understanding of hydrological

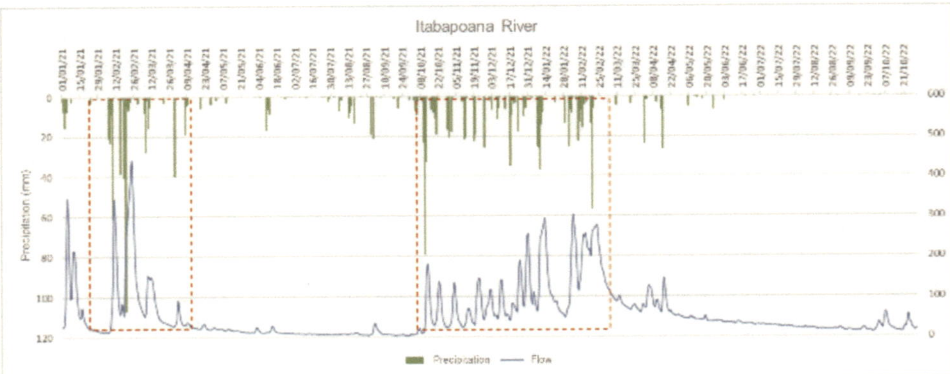

Figure 26.
Precipitation (mm) and flow (m³/s) records recorded at the 57836000 PCH Pedra do Garrafão downstream station, from 01/01/2021 to 10/31/2022. ☐ Significant records of precipitation (mm) and flow (m³/s) stand out.

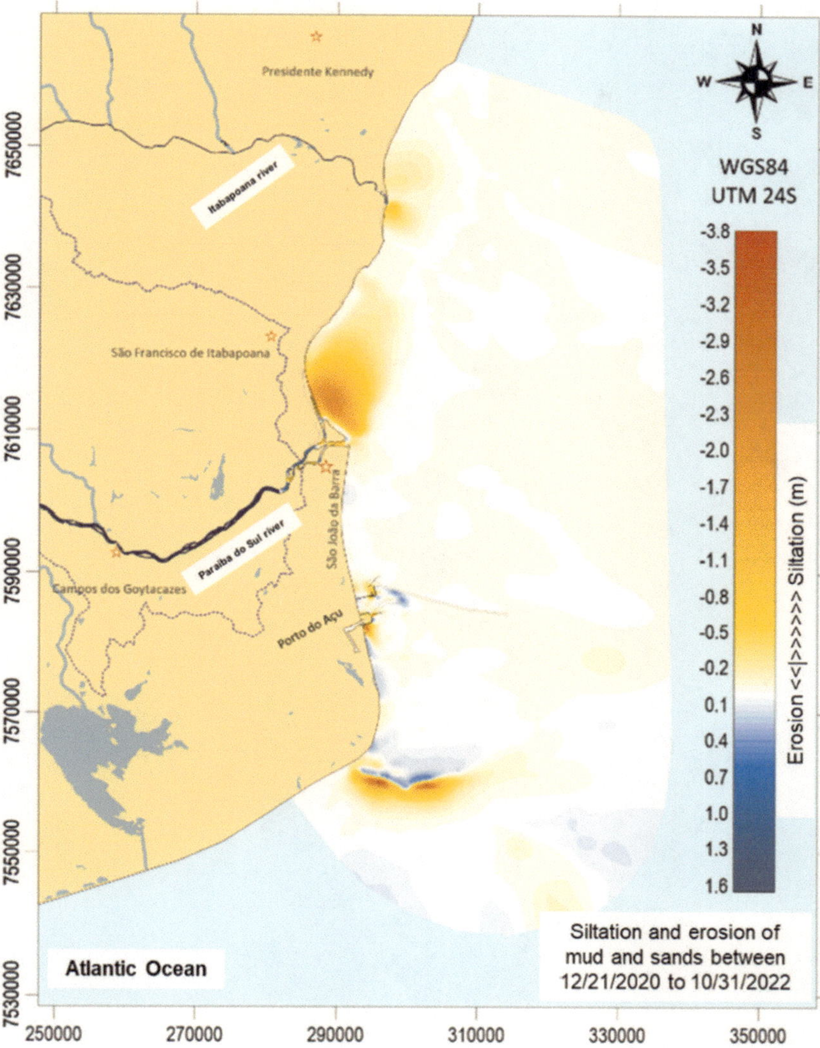

Figure 27.
Map of bathymetry alteration isolines due to silting and erosion from 12/21/2020 to 10/31/2022, considering existing sediments at the bottom and the TSS brought by the Paraíba do Sul and Itabapoana rivers. The T1 and T2 regions are zoomed in in the following figure. See siltation figures in T1 and T2 on the previous page.

dynamics and for the development of effective strategies for water resource management and disaster mitigation related to extreme precipitation events.

The results obtained by SisBaHiA offer a detailed view of the silting and erosion processes in the study area, covering the period from 12/21/2020 to 10/31/2022. It is possible to analyze sediment dynamics considering several classes of sediments, from TSS to carbonate gravel, with updates every 3 hours. This detailing capability allows for the production of specific maps, such as the one mentioned in **Figure 27**, making it easier to understand sedimentological changes over time and between different types of sediments.

The analysis of **Figure 27** highlights the detailed visualization of the silting and erosion processes in the studied area of Porto do Açu, using isolines to represent bathymetric variations. Negative values indicate depth loss due to siltation, while

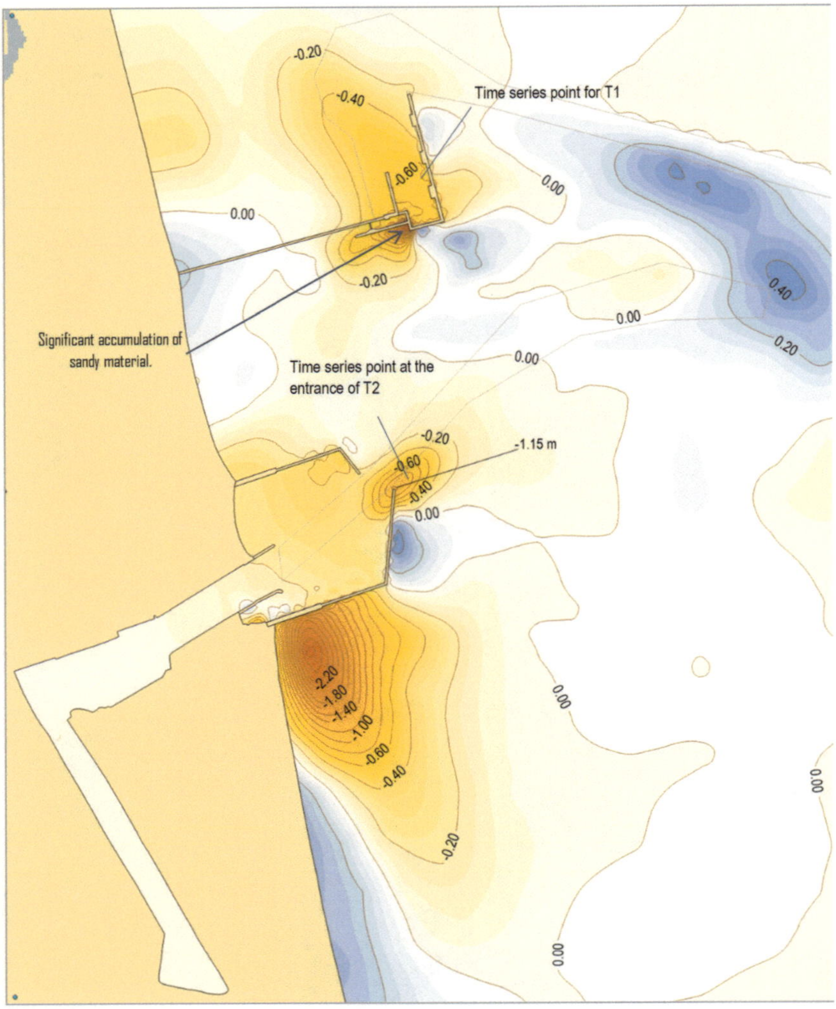

Figure 28.
Zoom in on the region of T1 above and T2 below, with detailed siltation and erosion isolines, including change values, according to the color scale of the general map in ***Figure 27****. Note the polygons delineated on T1 and T2, within which silted and eroded volumes were computed.*

positive values indicate depth gain resulting from erosive processes. This differentiation is important to understand sedimentological changes and their implications for the maintenance of navigability and the management of port infrastructures.

Figure 28 details the bathymetric variations in the areas of terminals T1 and T2, using colors to indicate erosion (blue tones) and siltation (brown tones). This allows for a visual understanding of depth changes, which is essential for assessing navigability and sedimentary management needs specific to each port region. At terminals T1 and T2 of the Port of Açu, there is significant siltation, with a predominance of fluid mud. In T1, about 676,000 m^3 of sediments were deposited, with 88% of fluid mud. In T2, silting totaled approximately 938,300 m^3, 76% of which was also fluid mud. These data indicate the need for specific strategies for sediment management and maintenance of the depth of navigation channels.

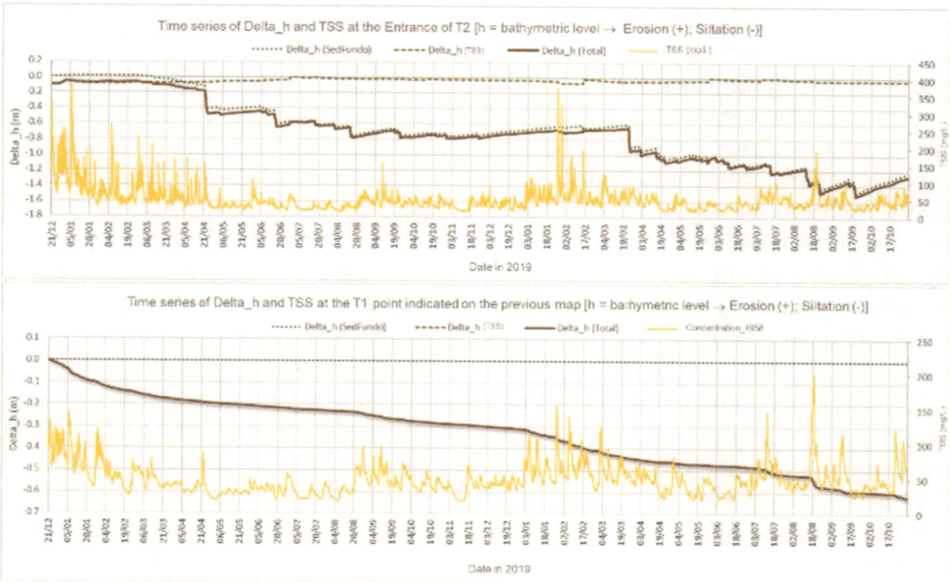

Figure 29.
*Time series of bathymetry variation, Delta h (m), and TSS concentration (mg/L), at the entrance of the jetties of T2 and in the internal area of T1, according to the points indicated in the zoom maps of **Figure 28**. Note that the scales of the axes of the graphs are different. Note that in the T1 area, the bathymetry variation with sandy sediments is practically nil; therefore, siltation with TSS is virtually equal to Delta-h-Total.*

The analysis in **Figure 29** reveals a direct correlation between the peak concentration of Total Suspended Sediments (TSS) and the increase in the siltation rate in the T1 and T2 areas. This indicates that TSS has a significant contribution to the silting process compared to sand, suggesting the need to monitor and manage these sediments to maintain the navigability and operational efficiency of the port.

4. Discussion

The analysis reveals a direct connection between water and sediment dynamics and port infrastructure management, showing how extreme events impact navigability and safety. The use of integrated models, such as SWAT and SisBaHiA, is essential to predict and minimize environmental impacts, corroborating the idea of their importance in the sustainable management of water resources and coastal zones.

The Port of Açu assesses climate-related risks and their impacts on operations, focusing on sea level changes, winds, waves, and precipitation. Forecasts indicate that climate change will have modest impacts on the port environment by the end of the century, but uncertainties grow for longer periods, emphasizing the need for continuous monitoring and updates [43].

The evaluation of the data between January 2021 and October 2022 identified significant temporal patterns in sediment movement and in the behavior of water flows in the Port of Açu, correlated with rainfall episodes. A clear link was observed between heavy rains and increased siltation and erosion, pointing to the vulnerability of port operations to such conditions. In addition, the hydrological response to precipitation was complex, with a time lag between heavy rainfall and the highest flows, evidencing the complexity of the processes within the watershed. These results agree with the literature on the importance of an integrated modeling approach in water management and climate change response, also suggesting the need for specific approaches to sediment management and climate adaptation in ports.

The influence of heavy rains and flow variations in rivers has a direct impact on port infrastructure due to siltation, highlighting dredging as an essential practice to preserve the depth of navigation channels and, consequently, the ability to receive ships of great draft. Effective sediment management, including regular dredging, is critical to maintaining port operations, taking into account the environmental effects of these activities [44].

The results reinforce the need for holistic mitigation approaches, integrating monitoring, modeling, and environmental management practices, offering valuable contributions to the literature for the development of public policies and adaptive management strategies. Sustainability in sediment management demands a synergy between the maintenance of navigation channels and the conservation of marine ecosystems, encouraging the reuse of sediments in ecological restoration and construction. The continued use of monitoring and predictive technologies reduces the need for dredging, decreasing the environmental impact [45].

Adaptation to climate change is essential for long-term port management, given the expected increase in extreme precipitation events. Implementing resilient strategies, improving hydrological forecasting and monitoring, and strengthening port infrastructures are vital to ensuring sustainable operations by mitigating adverse effects on port activities and nearby ecosystems [46–48].

The integration of hydrological and sedimentological modeling and meteorological data analysis are critical for informed decision-making, anticipating extreme events and their effects on port operations. This approach facilitates risk management by promoting port expansion in a sustainable and resilient manner [49, 50].

The practical implications of these results are important for stakeholders in the port industry and water resource management, underscoring the importance of proactive approaches to address siltation and water change, including:

- Modernization of port infrastructure to resist extreme events and mitigate siltation.

- Implementation of advanced monitoring and modeling systems.

- Formulation of public policies focused on sustainability and climate adaptation.

- Promotion of cross-sectoral collaboration for the development of integrated environmental strategies.

These recommendations are aimed at optimizing port operations, minimizing risks, and promoting sustainability.

5. Conclusions

The study achieved its objectives by illustrating the effectiveness of integrated modeling in the studies of hydrological and sedimentological processes, showing how this methodology can improve the accuracy of predictions and inform adaptive management strategies. The incorporation of advanced modeling techniques, which take into account variables ranging from meteorological to aspects of the physical geography of the region to oceanographic aspects, represents an advance, filling an important gap in the literature. This innovative approach provides an in-depth understanding of the impacts of heavy rainfall events on port areas and introduces methods that improve the prediction and mitigation of risks related to sediment and water dynamics, contributing significantly to the area of hydrology and water resource management.

The research identified a significant connection between extreme precipitation events and changes in sediment dynamics and water flows, with a particular focus on the Port of Açu. It was observed that heavy rains resulted in increased silting and modifications in water flows, compromising the efficiency of port operations. The application of integrated models, such as SWAT and SisBaHiA, has emerged as essential to analyzing and managing these effects. The results of this study enrich the field of integrated modeling of hydrological and sedimentological phenomena, underlining the need for comprehensive strategies to anticipate and reduce the impacts of rainfall events in port areas.

The results of the research are in line with the chosen theoretical framework, highlighting the relevance of integrated modeling for the understanding of hydrological and sedimentological systems. This study not only confirms the significant influence of extreme precipitation events on port areas, as theoretically predicted, but also extends existing knowledge by detailing the specific impacts and assessing the effectiveness of diverse mitigation strategies. Thus, the findings enrich the theoretical framework, illuminating the complexity of the phenomena studied and underlining the importance of integrated management strategies.

Limitations of the study may include the scale of data used, the representativeness of the modeled climatic conditions, and the applicability of the models in different geographic contexts. These limitations may influence the generalization of the results and the accuracy of the forecasts in different port areas. To overcome such limitations, future research could expand the database, include a wider range of climate scenarios, and test the applicability of the models in different environments. Acknowledging these limitations is important for the validity and continuous improvement of research in the field.

It is recommended for future investigations to explore models based on artificial intelligence (AI) and machine learning (ML) in the prediction of extreme precipitation events and in the analysis of their effects on sediment dynamics and water flows. Analysis of the interaction between global climate change and local water resource management strategies can reveal efficient adaptation approaches. In addition, the incorporation of interdisciplinary studies that combine socioeconomic factors with hydrological and sedimentological modeling could deepen the understanding of the challenges faced by port areas, contributing to the development of sustainable and integrated solutions.

Practical recommendations based on the results of the study include strategies for port resilience and sustainability, such as:

- Advanced modeling integration: adopt predictive modeling that encompasses meteorological, hydrological, and oceanographic variables to anticipate and mitigate climate impacts.

- Adaptive infrastructure development: build infrastructure that is aligned with the latest climate forecasts to ensure efficient and safe port operations.

- Enhanced monitoring: increase the resolution of environmental monitoring to collect detailed data that support predictive modeling.

- Future research: investigate the complex interaction among climatic and human factors and hydrosedimentological dynamics.

- Implementing innovative policies: create public policies that reflect the latest scientific insights for effective sediment management and environmental conservation.

- Adaptive management: adopt a proactive approach to port and environmental management, based on risk assessments and integrated modeling.

Among the challenges faced during the study, large-scale data collection and analysis and the modeling of complex systems are highlighted. However, the application of innovative methodologies, which incorporated the integration of multidisciplinary data, the use of advanced modeling technologies, and interdisciplinary collaboration, proved to be successful. The insights gained highlight the importance of methodological flexibility, adaptability to emerging findings, and the relevance of effectively communicating results to relevant stakeholders. These lessons are critical to enhance future research and practice in the field.

The research emphasizes the role of integrated modeling of hydrological and sedimentological processes in overcoming the challenges associated with extreme precipitation events in port areas. Advances in predictive modeling have been shown to offer significant contributions to water resource management and port infrastructure, providing valuable data for climate adaptation policymaking. This study represents a breakthrough in hydrological knowledge, encouraging the development of innovative solutions to address current and emerging challenges, and highlights the need for holistic and collaborative approaches in research and practical application.

Acknowledgements

We would like to express our gratitude to all those who have contributed in several ways to the realization of this work, whose support has been indispensable for the development of this book chapter. A special thanks is directed to the project participants and to the SisBaHiA Laboratory, as well as to the Coastal and Oceanographic Engineering Program of the Federal University of Rio de Janeiro, whose expertise and dedication significantly enriched this study. We extend our thanks to the managers of the Port of Açu, whose funding not only made this research possible, but also demonstrates their commitment to scientific advancement and environmental sustainability. The collaboration of everyone and institution was essential for us to achieve our goals and contribute to port resilience and environmental assessment for decision-making.

Author details

Lidiane dos Santos Lima[1]*, Paulo Cesar Colonna Rosman[2] and Juliane Castro Carneiro[3]

1 State Center for Monitoring and Alerts of Natural Disasters in the State of Rio de Janeiro, Federal University of Rio de Janeiro, Ocean Engineering Program, Rio de Janeiro, RJ, Brazil

2 Water Resources and Environmental Engineering Department, Federal University of Rio de Janeiro, Rio de Janeiro, RJ, Brazil

3 Porto do Açu Operations, Cais Açu Lab (Porto do Açu R&D and Innovation Program), Federal University of Rio de Janeiro, Rio de Janeiro, RJ, Brazil

*Address all correspondence to: lidianelima.ocn@gmail.com

References

[1] da Costa Sanches Galvão JR et al. Mobilization towards an integrated research network for studying runoff, sediment transport and climate of the Paraiba do Sul Basin. In: Frota MN et al., editors. Proceedings of the 1st International Conference on Water Energy Food and Sustainability (ICoWEFS 2021). Cham: Springer; 2021

[2] Gorodetskaya Y. Modelos hidrológicos híbridos para a bacia do rio Paraíba do Sul: acoplando redes neurais artificiais com transformada wavelet para previsão de vazão em curto prazo com ênfase na previsão de vazões extremas. Tese. s.l.: Universidade Federal de Juiz de Fora (UFJF); 2022

[3] Dunn RJK et al. Hydrodynamic and sediment modelling within a macro Tidal Estuary: Port Curtis Estuary, Australia. Journal of Marine Science and Engineering. 2015;**3**:720-744

[4] Guarnieri A et al. Effects of marine traffic on sediment erosion and accumulation in ports: A new model-based methodology. Ocean Science. 2021;**17**:411-430

[5] Izaguirre C et al. Climate change risk to global port operations. Nature Climate Change. 2021;**11**:14-20

[6] Mutombo K e ÖA. Towards port infrastructure adaptation: A global port climate risk analysis. WMU Journal of Maritime Affairs. 2017;**16**:161-173

[7] Janjić J, e Tadić, Lidija. Fields of application of SWAT hydrological model – A review. Earth. 2023;**4**(2):331-344

[8] Manisha KC et al. Application of SWAT hydrological model to simulate flow of Seti-Gandaki Basin. Jalawaayu. 2023;**3**(1):43-62

[9] Marmontel CVF et al. Aplicabilidade do Modelo Hidrológico SWAT Na Bacia Hidrográfica do Rio Paraibuna, SP - Brasil. IRRIGA. 2019;**24**:594-609

[10] Liping C, Yujun S e S. Monitoring and predicting land use and land cover changes using remote sensing and GIS techniques – A case study of a hilly area, Jiangle, China. PLoS ONE. 2018;**13**(7):1-23

[11] Darbra R, Maria R, Journée H, Wooldridge C. Environmental management tools for ports. 2011

[12] Pereira SP. Modelagem da Qualidade Bacteriológica das Águas Costeiras de Fortaleza (Nordeste do Brasil) [Tese, D. Sc]. 2012

[13] Ishikawa AK et al. Projeto Baías do Brasil - Porto do Açu, RJ. 2019

[14] Guannel G et al. The power of three: Coral Reefs, Seagrasses and Mangroves Protect Coastal regions and increase their resilience. PloS One. 2016;**11**(7):1-22

[15] Capozzoli C, Cardoso AO. Effects of land use change on discharge of the Paraíba do Sul river. Ciência e Natura. 2021;**42**(53):1-13

[16] Mukherjee A et al. Comparison of soil quality index using three methods. PLoS ONE. 2014;**9**(8):1-15

[17] Ribeiro AS et al. Reporting climate change impacts on Coastal Ports (NW Iberian Peninsula): A review of flooding extent. Journal of Marine Science and Engineering. 2023;**11**(3):477

[18] Shin ET et al. Integrated inundation modeling of flooded water in Coastal Cities. Applied Sciences. 2019; **9**(7):1313

[19] Maurya SP, Singh AKY. Agent-Based Modeling for Integrated Urban Water Management. Modeling and Simulation of Environmental Systems: A Computation Approach. 1st ed. CRC Press (Boca Raton); 2022

[20] SisBaHiA. Sistema Base de Hidrodinâmica Ambiental. SisBaHiA, COPPE|UFRJ [Online]. 2024. Available from: https://www.sisbahia.coppe.ufrj.br/

[21] SWAT. SWAT. Soil & Water Assessment Tool. [Online] 2024. Available from: https://swat.tamu.edu/

[22] Neitsch SL et al. Soil and Water Assessment Tool Theoretical Documentation Version 2009. College Station, Texas: Grassland, Soil and Water Research Laboratory, Agricultural Research Service and Blackland Research Center, Texas Agricultural Experiment Station; 2011

[23] NASA/METI/AIST/Japan Spacesystems and U.S./Japan ASTER Science Team. ASTER Global Digital Elevation Model V003 [Data set]. 2019. NASA EOSDIS Land Processes Distributed Active Archive Center

[24] INDE. Infraestrutura Nacional de Dados Espaciais. Portal brasileiro de dados geoespaciais – SIG Brasil. [Online] 2022. Available from: https://visualizador.inde.gov.br/

[25] Embrapa Solos. Correlação com WRB/FAO e Soil Taxonomy. Sistema Brasileiro de Classificação de Solos (SiBCS). [Online] 2022. Available from: https://www.embrapa.br/solos/sibcs/correlacao-com-wrb-fao-e-soil-taxonomy

[26] POWER Data Access Viewer. NASA POWER. NASA Prediction of Worldwide Energy Resources. [Online] 2022. Available from: https://power.larc.nasa.gov/data-access-viewer/

[27] NASA. POWER Data Methodology. NASA POWER | Docs. [Online] 2022. Available from: https://power.larc.nasa.gov/docs/methodology/

[28] NASA POWER. POWER Docs. NASA POWER | Docs. [Online] 2022. Available from: https://power.larc.nasa.gov/docs/

[29] Rosman PCC. Referência técnica do SisBaHiA. Rio de Janeiro: Laboratório SisBaHiA, Programa de Engenharia Oceânica da COPPE, Universidade Federal do Rio de Janeiro; 2024

[30] Figueiredo AG et al. Geomorfologia e sedimentologia da plataforma continental. In: Kowsmann RO, editor. Geologia e Geomorfologia. Vol. 1. 2015. pp. 13-32

[31] AVISO+. FES2014. FES (Finite Element Solution)- Global tide. [Online] 2016. Available from: https://www.aviso.altimetry.fr/en/data/products/auxiliary-products/global-tide-fes/description-fes2014.html

[32] HYCOM. HYbrid Coordinate Ocean Model (HYCOM). Center for Ocean-Atmospheric Prediction Studies (COAPS). [Online] 2022. Available from: https://www.hycom.org/

[33] Hersbach H et al. ERA5 hourly data on single levels from 1940 to present. Copernicus Climate Change Service (C3S) Climate Data Store (CDS). 2023

[34] Refsgaard JC. Parameterisation, calibration and validation of distributed hydrological models. Journal of Hydrology. 1997;**198**(1–4):69-97

[35] McMillan HK, Booker DJ, e Cattoën, C. Validation of a national hydrological

model. Journal of Hydrology. Part B. 2016;**541**:800-815

[36] SWAT. SWAT-CUP. Soil & Water Assessment Tool. [Online] 2024. Available from: https://swat.tamu.edu/software/swat-cup/

[37] Abbaspour KC. SWAT-CUP, SWATCalibration and Uncertainty Programs. 2015

[38] Hora MAGM. Diagnóstico preliminar das condições hidrossedimentológicas do rio Paraíba do Sul e seus principais afluentes. Água em Revista. 1998;**10**:27-35

[39] Moriasi DN et al. Model evaluation guidelines for systematic quantification of accuracy in watershed simulations. American Society of Agricultural and Biological Engineers. 2007, 900;**50**(3):885

[40] Porto do Açu Operações. Dados dos níveis de maré de 21/12/2020 a 31/10/2022. Estação T1 – Ponte de Acesso. 2022

[41] INMET. Eventos extremos de fevereiro de 2022 no Brasil. Brasília, DF: IMMET; 2022

[42] INMET. Alerta conjunto chuvas intensas região Sudeste. Brasília, DF: IMMET; 2021

[43] Porto do Açu Operações S.A. Porto do Açu: Avaliação dos riscos climáticos – Resumo Executivo. Rio de Janeiro: Porto do Açu; 2022

[44] Delanoy RA, Díaz-Asencio M, Méndez-Tejeda R. Effect of extreme weather events on the sedimentation of the Bay of Samaná, Dominican Republic (1900–2016). Journal of Geography and Geology. 2019;**11**(3):56-73

[45] Pellegrini M et al. Sustainable sediment management in coastal infrastructures through an innovative technology: Preliminary results of the MARINAPLAN PLUS LIFE project. Journal of Soils and Sediments. 2020;**20**: 2685-2696

[46] Loza P, Veloso-Gomes F. Literature review on incorporating climate change adaptation measures in the design of New Ports and Other Maritime Projects. Sustainability. 2023;**15**(5):4569

[47] Devendran L et al. Adapting to the impacts of climate change on port operation. Australian Journal of Maritime & Ocean Affairs. 2023;**15**(2): 107-126

[48] Zheng S et al. Seaport adaptation to climate change disasters: Subsidy policy vs. adaptation sharing under minimum requirement. Transportation Research Part E: Logistics and Transportation Review. 2021;**155**:102488

[49] Omran ES, Dandrawy ME. Integrated hydrological modeling and geoinformatics for harvesting and simulating Mountain Torrents on the Area Stretching Between Port Sudan and Ras Bennas, Red Sea. In: Omran ES, Negm AM, editors. Egypt's Strategy to Meet the Sustainable Development Goals and Agenda 2030: Researchers' Contributions. Cham: Springer; 2022

[50] Merico E, Dinoi A, Contini D. Development of an integrated modelling-measurement system for near-real-time estimates of harbour activity impact to atmospheric pollution in coastal cities. Transportation Research Part D: Transport and Environment. 2019;**73**:108-119

Chapter 5

Towards Integrated and Sustainable Water Management in Water-Scarce Arid Environments: Case of Sudan

Abdelbagi Y.F. Adam and Mohamed B.O. Osman

Abstract

In water-scarce arid region, integrated strategy for water resources management is critically needed to cope with water scarcity and the ever increasing demand and pressure on water resources induced by lessened rains, warming climate tendencies, population growth, and environmental sustainability considerations. In Sudan, despite the "relative" abundance of multi-sourced water quantities, yet the spatiotemporal variations adversely affect water availability and accessibility to the population majority in various arid and semiarid parts of country. Water deficit hampers achieving the socioeconomic development, especially in the concurrent climate change, large-scale transboundary river damming, conflicts over water resources, growing population, and economy contexts. Efficiently addressing of these issues and enhancing water resources management strategies are imperative necessities. Integrated water resources management (IWRM) approaches and GIS techniques can together improve decision-making and sustainable management strategies in these changing climatic and hydrological systems. IWRM as inclusive framework and participatory approach while GIS represents spatial analysis tool for water resource monitoring, vulnerability assessment, planning for addressing demands, and adaptation alongside accounting for population growth and climate change. This chapter discusses the state of water resources, efforts for efficient management, and prospects of enhancement in Sudan, to attain sustainability amidst the recently changing physical and anthropogenic environments.

Keywords: water scarcity, arid/semiarid regions, IWRM, GIS, water resources, climate change

1. Introduction

Water management is a globally challenging issue that threatens various of sectors and multi-leveled users. In spite of its renewability, water resources are recently being researched and increasingly questioned in terms of their quantitative and qualitative availability over space and time by several groups of users at various uses and fields, for instance, households, agriculture, and industries [1].

Water resources management currently confronts an increasing burden and excessive pressure induced by many factors, to mention some, global warming, global population growth, urbanization, and the ever-growing economy and industries. Water management, thus, does not only encounter the quantitative and qualitative adequacy challenges; the spatiotemporal obtainability, the sustainable and ecological-friendly water utilization, is required as well [2].

This was because of the crucial and vital roles water resources play for each field of life and human life activities [3, 4]. Besides the importance of water for the different life aspects and activities, water management issues and their relevant research have gained a rising attention among the researchers of various professionals, science, and engineering disciplines such as geography, civil engineering, environmental sciences, and others.

Communities and nations, likewise, organize themselves in relation to water and develop institutions and practices to manage it [5]. As several of endangering and adversely affecting natural and human factors confront effective water resources management and threaten life and economy. The warming tendencies and lessened and variable rainfalls induced by climate change [6], population growth, and its accompanying activities such as urbanization, and industrialization, justify for the importance of studying water resources and the increasing attention paid to consistently tracking of its quantities and qualities since the ancient ages up to this moment of history.

Since time immemorial, human societies grew and prospered around rivers and freshwater courses, and several of ancient civilizations prevailed and thrived then failed according to their success and failure in agriculture and irrigation systems, which was indicating the stability and economic power of state. It is obvious that water is not only related and necessary to all different life aspects and activities. But it is a main determinant factor for the state to sustain and a crucial element shaping its socioeconomic profile and cultural aspects [5, 7, 8].

Over thousands of years, investments and innovations in water management have contributed to socioeconomic development and substantial civilizational achievements [7, 8]. This contribution has been instrumental in providing water supplies for drinking and irrigated agriculture, facilitating the attainment of these accomplishments [9]. There are intricate connections of water with the sectors of food, energy, health, and community development.

This crucial role of water in the various sectors highlights the importance of its sustainable management for achieving the overall sustainable development [2, 10]. Additionally, it highlights the necessity of incorporating water security considerations into policies and methodologies to ensure the preservation of water resources for the well-being and prosperity of present and future generations [9, 10].

Taking a lesson from history, as Sudan possesses considerable water resource base, developing these water resources and properly managing and exploiting them offers a tremendous opportunity for economic development, social viability and eventually establishing a strong and developed country such that political stability is achieved.

Therefore, this chapter attempts to highlight how GIS and IWRM approaches can be used as spatial analysis tool for water resource monitoring, vulnerability, assessment, and planning for utilization and adaptation in Sudan. To improve decision-making and management strategies in the changing hydrological and climatic patterns. This will help to properly manage water demands, account for population growth and climate change, and ensure that water is used efficiently and sustainably. Hopefully, this will be a slight contribution in pushing the wheel off development in Sudan forward.

DOI: http://dx.doi.org/10.5772/intechopen.1004717

2. Sudan's hydro-geographic and climatic settings

Hydrologically, Sudan water resources are generally divided into two major sources: Nile and non-Nile sources. The Nile source is the river Nile and its tributaries rivers that flow in Sudan. Non-Nile sources of water are mainly represented on the groundwater, seasonal running rivers, and valleys "wadies" and the rains water.

Proper understanding of the state of water resources and planning for their management and development require a solid knowledge of the physical and human geographic contexts of the country. Geography and geographers, on the other hand, have an interest in studying water from both environmental and anthropogenic perspectives and implications. Because there are relationships among water and the sectors of food, energy, health, and community development. The role water plays in all of those sectors is significant and indispensable.

For this, the quest for water security, preserving water rights of future generations and setting approaches and policies for effective water management, should bear into consideration the geographic knowledge, as a basis to and understand water phenomenon and how it affects and being affected by the urban environment, physical and biological ecosystem components, as well as the socioeconomic traits. Geographic methods manipulate all of these variables properly and equally. This knowledge of geographic settings and extent will facilitate well delineation of water characteristics and potentialities in Sudan for purposes of planning and formulating development strategies.

Sudan is located in the northeast of the African continent between 8°–22°N, 21°–39°E, surrounded by Egypt from the north, and on a counter-clockwise basis comes Libya, Chad, Central Africa, South Sudan, Ethiopia, Eritrea in the north, while the Red Sea separates it from Saudi Arabia. Look at **Figure 1**.

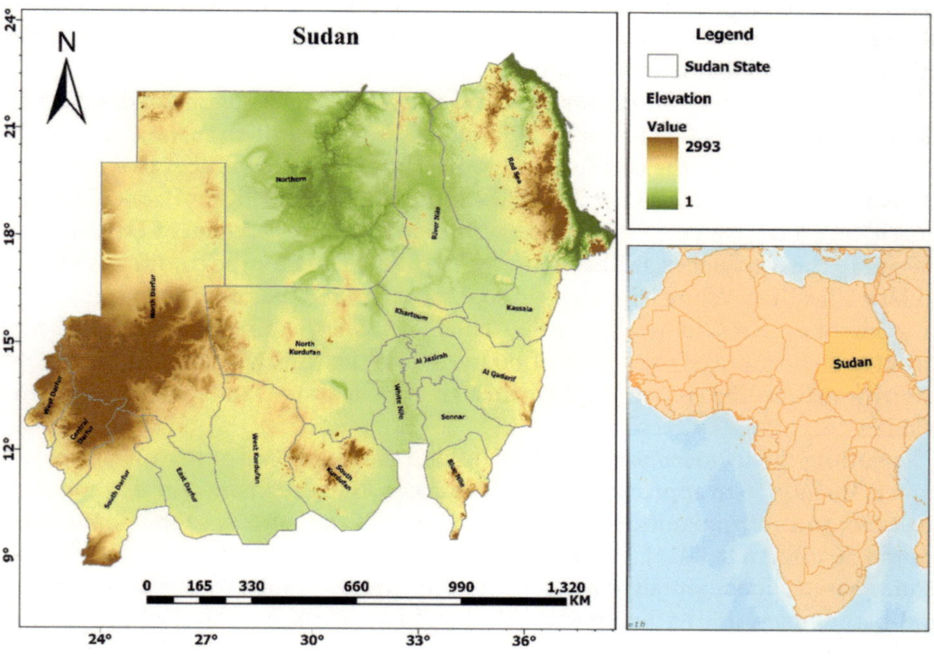

Figure 1.
Previews location and some geographic aspects of Sudan.

This location refers to the fact that country is situated in the tropics, and the predominant climate of Sudan is a continental climate, except in the regions of Red Sea surroundings and other mountainous and highlands. However, the country is laying over extensive and varied tropical lands, where several climatic and ecological zones prevail and vary south-northward from high rainfall savanna lands to the hyper-arid Sahara desert. While semiarid and low-rainfall savanna lands prevail in the central regions.

The inherited continental climate of the so-called intertropical convergence zone, "ITCZ", gives causes of spatial and quantitative variations of the Sudan rains; temporal and seasonal disparity is a further defining characteristic of Sudan rains. For this, rainfall patterns in Sudan vary annually and geographically, depending on ICTZ, extreme northern region receives almost no rain, central regions receive moderate rates of 200–700 mm, while the southern part rains exceed 1500 mm.

These rains, known as "the monsoon" that precipitate seasonally (June-September), are the Sudan's most prevailing rainfall; it is very critical climate variable for greater part of Sudan and majority of it is population as it plays a significant role in major economic and productive activities as the essential water source for people, animals, agriculture, and natural vegetation, respectively.

The geologic map of Sudan denotes that the country is formed into several geologic formations that contain aquifers. Therefore, groundwater is potentially available in large areas of the country. Major aquifers cover more than 50% of the area of the country, and they fall into: The Nubian sandstone aquifer, the Um Ruwaba formation, the recent alluvial Wadi aquifer, and the hard rock aquifers of the basement complex [11]. Studies indicate for plentiful groundwater amounts in Sudan but not agreed on an exact estimation.

Topographically, The Nile River system, as the dominant physical feature, penetrates the country from extreme south to the northern border in extended plains that gently slope to the north and disturbed with few inselbergs and some other limited mountainous highlands, such as "Jabel Marra" and Red Sea mountains. Nile is not only the main factor shaping the principal geomorphological and topographic aspects of the country, but it is also the main drainage system of the country.

Throughout much of the country, most of wadies and rivers of Sudan drain toward Nile basin, that is, into Nile or one of its tributaries, some of them succeed to reach the Nile while others lost in the surrounding plains [12]. However, number of watercourses of Sudan drainage system do not flow toward the Nile, such as "Khor Baraka" and "Alqash" seasonal rivers and some other wadies in the western region of Darfur.

Prior to the confluence at Khartoum, Nile tributaries enter the country as white and blue Niles. The White Nile starts in southern part and maintain an extremely low gradient until it is joined by the Blue Nile. The Blue Nile rises in the Ethiopian Plateau and contributes much of the floodwaters of the Nile. Nile then keeps flowing in northward curving course [12] to form a great river system and major water source in Sudan.

Regarding the socioeconomic profile, despite the rapid pace of urbanization (from 8.3% in 1956 to approximately 40% in 2015), Sudan remains rural in social, economic, and cultural outlooks with the majority of the country's total population (around 36 million in 2014) living in rural areas and pursuing extractive livelihood systems based fundamentally on the direct utilization of land and natural resources. Land-based renewable natural resources, therefore, are the backbone of the principal sectors of the Sudan economy [13]. The major economic activities in the country are agriculture and grazing. Another sector dependent on rainfall is gum Arabic production, of which Sudan is the world's leading producer [14].

Furthermore, the diversity of environmental conditions, especially in relation to water availability, rainfall amount, and soil type has given rise to a wide variety of habitat and livelihood options. As a typical African Sahel, on Sudan's central rain lands that extend from west to east, water is a main limiting factor to development in addition to local environmental conditions and technical and marketing constraints; for the reason that resources management, land use system, economic activities, and human adaptation were concentered on traditional rain-fed cultivation and animals herding. These processes have affected cultural and social landscapes as well as the economic relationships amongst the population groups [13].

Other minor employment and livelihood means are witnessed in Sudan, nevertheless the above-mentioned form the holistic socioeconomic aspect of the country and prevail the most. Obviously, all are directly related to water resources. For this, the contemporary geographic, hydrologic, and socioeconomic contexts of Sudan waters urge an immediate intervention and science-based action from the government of Sudan to protect the country's population and economy from the threats of water resources.

3. Water resources between security and scarcity

The state of water viability in a region is judged by the quantitative and qualitative availability of water. Water is usually addressed and evaluated using various criteria that are commonly expressed about through certain terms referring to the quantitative state of water such as water security, water scarcity, water stress, and others. However, there furthermore specialized and technical terminologies to highlight the state of the water.

Until recently, the discussion of water issues occurs most commonly within the frame of water security and scarcity. This is generally used to express the overall state of water availability for the needs of the population in a particular location. However, with developments of very specialized scientific domains and the technological advancements, water security/scarcity description has undergone through very technical and specified quantification and measurements using mathematical modeling, according to the discipline of study.

For this, water security/scarcity concepts and standards are now being examined and debated at different levels, ranging from global and regional and end with household and individuals' levels of water security. Moreover, it is common that water security and scarcity are defined, explored, and discussed within specialized fields such as agriculture, public health, and food security in a customized standards and approaches.

Water security is defined as the provision of acceptable quantity and quality of water for health, livelihoods, ecosystems, and production in conjunction with a level of water-related risks and their impact on the environment and economy [15]. According to the UN, water security is the ability of people to sustainably maintain access to sufficient quantities of water of acceptable quality, to maintain livelihoods, human well-being, and social and economic development, to ensure protection from waterborne pollution and disasters related to water, and to preserve ecosystems in a climate of peace and political stability [16].

For this, one can say that water security is generally achieved through sustainable management and efficient use, which prevents lessening of the quantitative and qualitative availability of water and reduces the risks facing water in general.

Water scarcity, on the other hand, can be defined as the state of general shortage that a region suffers from in its sustainable water resources. The state of water scarcity is generally exacerbated by several climatic factors, reliance on a shared water source, and increased water losses due to the increase in population, consumption, and evaporation [9]. While water resources endure limited and not increasing and developed, steady in favorable conditions. It goes without saying there is a state of water scarcity when demand increases, while available water resources are limited, or exceeding sustainability limits.

4. Water resources in Sudan

Sudan is among the least developing countries that is situated in the arid regions and populated with rural communities, in majority. The population of Sudan predominantly relies on extractive and primary economic activities. Therefore, they will be significantly affected by the environmental, climatic, and economic factors that are caused and affected by the water deficit as economy and livelihood means are directly interrelated to the water.

Therefore, it goes without saying, climate and environmental change, poor governance, and socioeconomic factors are among many causes and effects for water scarcity in Sudan [17] and eventually led to adverse consequences, such as, the poverty, food insecurity, poor and inefficient water management systems, and others.

Well-studying, documenting, accurate measurements, and monitoring of the state of water resources are key prerequisites for improved utilization and efficient management that based on informed policies [9]. For this, it is worthwhile and important to delineate the state of water resources and supplies in Sudan, identify their challenges, and highlight possible improvement interventions.

4.1 The current state

The River Nile and its tributaries are the primary sources of water in Sudan, accounting for 73% of the annual freshwater [18]. According to the REDD+ estimations, Sudan's total natural renewable water resources are estimated to be 89 km^3/year, including 20% from rainfall and the remaining 80% flowing in from the origins of transboundary upstream. It worth mentioning that, Sudan is utilizing about 14.6 billion cubic meters "BCM^3" of its 18.5 BCM^3 share of the 1959 Nile waters agreement. The overwhelming part (96.7%) goes to agriculture, domestic household withdrawals, and industrial sectors amount to 2.6 and 0.7%, respectively [13].

Regardless the Sudan share from Nile and the actually utilized quantity, the reliance on externally generated surface waters is a key feature of Sudan's water resources, and this is critically important to economic development in the country as the flow of waters is highly variable on the annual basis and subject to long term regional conditions due to political, environmental, and climatic changes. This marks a reduction susceptibility of Nile as main water source for Sudan.

In addition to Nile Sudan experiences incidence of other water sources such as rainfall, surface flow, and groundwater. However, Sudan water resources are relatively limited and disturbed by the uneven quantitative distribution of that water regionally and seasonally. In unfavorable climatic conditions, rainfall received had showed variability and large deviations from the recorded and projected mean annual rainfall.

The greater part of the country lies within the arid and semiarid zones of high evaporation rates and low rainfall, approximately, less than 300 mm annually. The total amount of freshwater from internal and external sources is approximately

30 × 109 m3/year, bringing the per capita water availability below the water stress limit of 1000 m3 [19]. This was further exacerbated by the fact life and economy in Sudan revolve around water.

In spite of its various water sources, generally, Sudan is suffering from the insufficiency in water resources, and in overall national scale, the water balance is negative. According to the National Water Corporation, the water demand, 32.1 km^3 per year, has already exceeded the country's total water resources of 29.5 to 31.5 km^3 per year.

Furthermore, the expanding national population with a 3.2% annual growth rate puts further strain on water balance. For this, with an annual availability of 500–1000 cubic meters per year for individual, the ESCWA study assessed water availability per capita for Sudan population, using the abundance/scarcity scale, the study concluded that Sudan, in overall, experiences a state of water scarcity, regardless of local and environmental variations in each region or locality [20]. The below figure indicates the accessibility of locals at different regions over the Sudan (see **Figure 2**).

Owing to this weakened management in developing arid countries, such as Sudan, there is always a distinct rainy season, followed by a dry period for the rest of the year. During summer, precipitation occurs mostly in convective rainstorms that are typically scarce and of short duration, but sometimes can be very intense. This, at times, causes massive water runoff and results in devastating floods [22].

By looking at water data and the corresponding population taking into consideration the steadily growing economy and population size. Accordingly, in the following figures, intersections of rain quantities and distribution as the main source for water in greater parts of Sudan, the incident temperature, and population distribution give general indicators to drought index and water availability over the country (**Figures 3–5**).

From the above overview, despite the substantial freshwater resource base, that is, almost half of the Nile Basin is found within the country, considerable rainfall amounts and groundwater aquifers are globally renowned, Nubian Sandstone Aquifer in particular. However, the poor management, very basic infrastructure, and undeveloped groundwater reserves had hindered the achievement of water security and bridging the acute disparity of inter-regional water availability and remarkable spatiotemporal fluctuations. One can say Sudan has tremendous water potentials and potentialities with poor and inefficient exploitation.

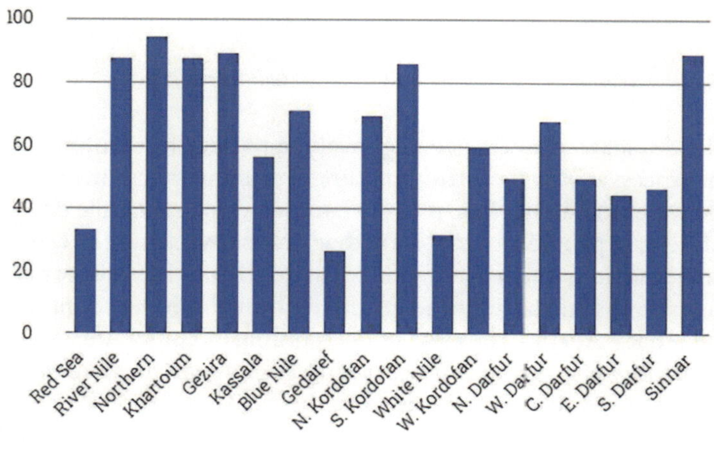

Figure 2.
Shows the accessibility rate to fresh water for each state citizens. Source [21].

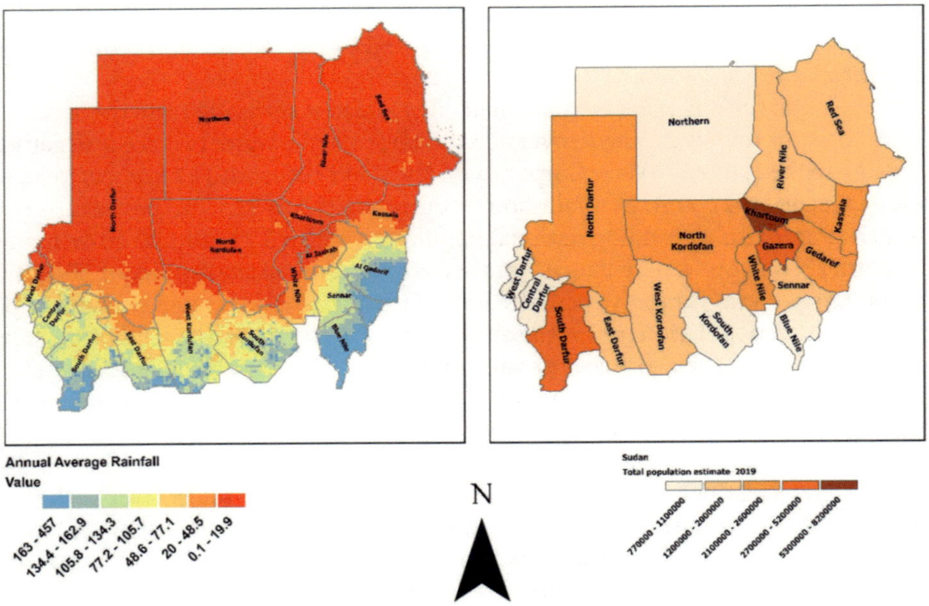

Figure 3.
View the contrasts the distribution of rains and population over Sudan. Source: Prepared by the authors.

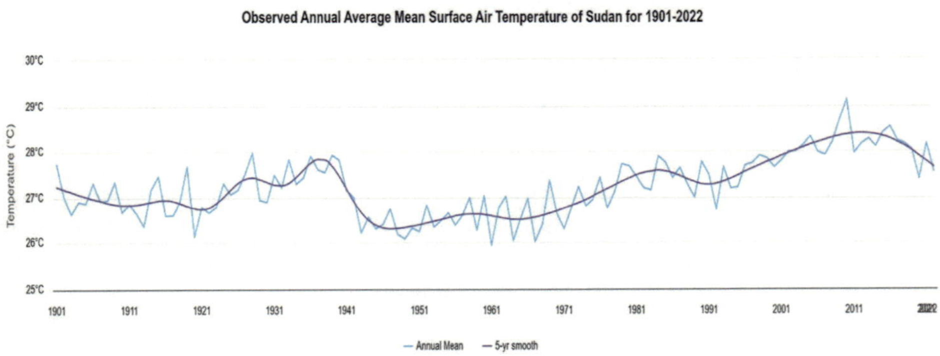

Figure 4.
Shows the trends of annual mean temperature of Sudan (1901–2022). Source: [23].

Obviously, this state of water had negatively impacted the planning, design, and construction processes of water infrastructure development and management [22]. Likewise, the water imbalance has not only hampered the economic development and adversely influenced people's lives, it is further a causative source of economic hardships in the drier regions and a driving force for conflict in the country [13] owing to the predominance of rural society and economy throughout the country.

To alleviate that hardship, the clear understanding of constraints and opportunities of the available water resource base, the integrated planning and development, good governance, appropriate institutional framework, and proper policies and regulations, are all prerequisites for achieving national goals of the improved water management, hence, sustainability, poverty reduction, economic growth, resilience and better adaptation to disasters induced by climate and water scarcity [24].

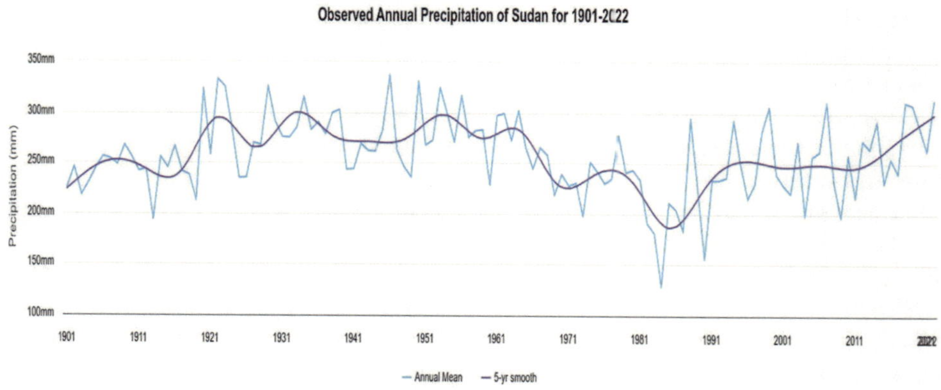

Figure 5.
Displays the trends of annual mean of rain incidents in Sudan (1901–2022). Source: [23].

4.2 The encountered challenges

The state of Sudan water resources, with its all of different sources and forms, can be characterized by "the uncertainty". All of the Sudan water sources are shrouded in uncertainties as no adequate and detailed studies covered them, and also there is no information system or national database that stores data and constantly monitors change over time, except of few and scattered studies.

Accordingly, Sudan possesses a various array of water assets along with rainfall, groundwater, and nonconventional water resources along with wastewater reuse and desalination. However, every single water source encounters one or more threatening challenge, whether natural or anthropogenic, which complicate sustainability of that source and hinder the improved utilization and development opportunities.

In the endeavor of sustainable exploitation and management, rains and surface water, Nile and non-Nile resources, confront many challenging threats on both earth surface and atmosphere; to mention some, not limited to transboundary damming, warming tendencies, climate change-induced lessened and variable rain patterns, in addition to the spatiotemporal variations of water resources that influenced the water supplies and accessibility, especially in arid Sudan rain lands [18, 25].

Recent climate projections indicate erratic seasonal rainfall, greater inter-annual variability, and more frequent and intense droughts will prevail in Sudan [26]. These all are very challenging as these are not subject to full human control, and their variability is unpredictable, at times to the water policymakers. However, modeling of scenarios, strategic planning, strategic adaptation measures, and preparedness, will significantly aid enhancing water management.

Uncoordinated transboundary damming, uncontrolled floods, the lack of the data and studies for planning, technical and technological incapability, poor infrastructures, lack of complementary and profitable projects associated with water resources, absence of strategic plans to address the threats, and development of surface water resources clearly indicate the weakened institutional water governance.

In the same regard, groundwater is not better and similarly confronted by the abovementioned challenges and further ones, that is, one of the main difficulties in groundwater management is that differently from surface water, groundwater is not directly visible to the decision-makers, its response is usually very slow, and the consequences of any management policy are not immediate and could take a long time before they are realized. However, despite they are the essential prerequisites for any successful assessment, planning, design, operation, and management of groundwater systems. Sudan faces critical challenges of lack in of quantitative and qualitative monitoring and scarcity of information on aquifers geometry and their hydrogeological characteristics [11].

In spite of the considerable number of previous studies that have been conducted on groundwater resources in Sudan, there is no unified and exact value of their estimated strategic potential. However, the various literature estimates refer to the hugely abundant groundwater quantities. Regrettably, the proper planning and decision-making for Sudan's groundwater utilization is still challenging and needs further detailed and specialized studies. Nevertheless, proper planning and decision-making for Sudan's groundwater utilization is still challenging and needs further detailed and specialized studies. Groundwater monitoring and water information system in Sudan have been very minimal and suffered huge deterioration over the years as a result of inadequate finance and capacity [11].

Categories	Groundwater	Surface water	Atmospheric water
Natural and environmental challenges	• Pollution and mineral dissolving • Salinity	• Limited storage vessels • Uncontrolled floods • Evaporation • High sediment rate and Water Hyacinth	• Warming climate • Regional and seasonal disparity
Socioeconomic challenges	• Population growth • Pollution • Changing water use patterns	• Population growth • Pollution • Rapid economic growth • Changing water use patterns	Lack of water harvesting
Political and administrative challenges	• Political instability • Shared aquifers • No coordination between the countries	• Political instability	Regional and seasonal disparity
Inter-regional and transboundary challenges	• Unbalanced Development • Conflicts over aquifers • Transboundary basins	• Down-upstream Conflicts • Uncoordinated river damming • Unbalanced regional development competitiveness	
Technical challenges	• Not easy to manage • Absence of lanning and strategic agenda • Lack of data for planning • Inadequate investments and finance in water sector	• Lack of DB and ISa • Poor infrastructure • Lack of effective harvesting projects • Poor river development • Lack of data for planning	• Uncertainties • Insufficient preparedness and adaptive measures • Lack of data for planning

Table 1.
Summarizes the challenges hindering achieving water security in Sudan.

International coordination for equitable use and development is another impeding factor to sound management of Sudan groundwater. Sudan shares several aquifers with neighboring countries, this includes, the Nubian Sandstone Aquifer with Egypt, Libya, and Chad, and the Um Ruwaba Aquifer shared with South Sudan. Nevertheless, there is not yet a regional institutional arrangement for the sharing of water from this aquifer [11]. The below table summarizes and highlights the challenges facing the various water source management, see **Table 1**.

For this, an urgent intervention and more enhancements should take place. Sustainability measures and policies should be made along with the studies and strategies for developing these water resources. All based on detailed and ad hoc studies and prescription of the challenges and threats of the water resources sustainability.

5. Paving for water resource enhanced and sustainable management: GIS and IWRM roles

The levels of planning for water management vary in scale from international to local [9]. This chapter concentrates on Sudan specifically, for this the discussion of water security on this chapter will be confined on the national level of water security.

One of the principal roles of national-level water security planning is to concentrate on utilizing water resources to achieve the economic and social development for different communities. For this, it is vital to involve balancing technical and administrative requirements for the water sector within the dynamic political landscape to realize overarching interests. Adopting a comprehensive and integrated participatory approach by involving all stakeholders in planning and implementation and then clearly defining the strategic goals of water security are imperative necessities.

The absence of precise scientific analysis, continuous and regular water resources inventories, population census, agricultural and industrial inventories, and proper estimation of the needs of each sector collectively and specifically, poses major challenges to sound planning based on scientific foundations that incorporate accurate data and robust knowledge of the reality of water resources in Sudan.

This is combined with the currently international heated competition and conflicts over water resources on the international level amid the concurrent technological and scientific capabilities leading the massive transformation in perspectives of water security and various national security in general, for example, mega dams and securing substantial expansion in agriculture and industry that consume ample water amounts.

The lack of monitoring, surveillance, and associated infrastructure are among the primary challenges facing the water sector. It further leads to poorly diverse and accumulated data required for the planning and informed decision-making. To bridge this gap, the use of modern monitoring and surveillance systems technologies will provide plentiful data. Utilization that numerous data sources and their diverse applications and modern data analytics and modeling capabilities will enable having integrated and inclusive perspectives for various field to decision-making that considers the various fields such as agriculture, population, and hydrology, to achieve a balanced development agenda *via* sustainable water resources management.

The government officials and academics should conduct a wide and precise field inventory to determine the quantitative and qualitative parameters necessary to achieve water security and meet the requirements for economic and social development. Additionally, there should be continuous planning and efforts to conserve and develop necessary water resources through sustainable and innovative solutions. Innovations in water management sector should work on finding integrated and complementary water resources management, through enhancing the resources resilience amid context of pressing factors and limited resources, to achieve and endure water security.

In short, with special reference to Sudan, increasing the political will, using the recent technological solutions, following more participatory approach, and reforming technical and institutional support, considerably contribute to reaching the state of water security and stabilizing its relevant socioeconomic consequences to foster development.

Water resources management is a complex issue that encompasses various tasks and activities with different spatial and temporal scopes. The complexity of tasks increases when transboundary water issues arise, introducing additional factors such as diverse legislation, policies, and information system frameworks in countries sharing the watershed. Designing, building, and operating water information systems are crucial for the successful management of water resources throughout all stages, including assessment, planning, design, operation, and management.

Water information systems can serve both short term and strategic tasks for water resources management, through providing real-time information on the location, availability, quality, and current uses of local and regional water resources data, thereby enhancing sustainable water management and knowledge-based decision-making. Therefore, in the framework of paving for water resource enhanced and sustainable management, GIS and IWRM have critically important and revolutionary roles that will enrich the water management and pave the road toward sustainable water management, as explained in following:

5.1 GIS roles

GIS is a powerful tool that assists, with major transformations, all of the natural resources management kinds. The development of GIS has empowered practitioners and mangers to better understand land, its dynamics, and how resources such as water interact with it. In relation to sustainable agriculture, water management, and environmental conservation, this understanding is useful.

GIS plays an essential role in the management of water resources. It helps visualizing and verifying spatial data related to water sources, including water quality, water quantity, and water distribution. Water sources can be monitored, climate change vulnerability can be assessed, and modeling techniques can be advanced using GIS. Additionally, it can be used to identify areas at risk of water shortage and analyze the impact of human activities on water resources [27].

GIS can help improve decision-making and management strategies by providing accurate and up-to-date information on water resources [28]. GIS is utilized to delineate the sub-catchment in hydrological modeling analysis, and it is convenient to calculate runoff at the outlet. GIS techniques can be used to evaluate the flow characteristics of rivers and streams in arid climates [29]. It can also be used to create maps and models that can be used to predict future water availability and to plan for future water demands [30].

In addition to the aforementioned advantages, in Sudan, as water data is typically collected by government institutions, lacks consistency and spatiotemporal coverage, and changes in water management approaches and practices affect the quantity and quality of the required data and information. Water management has become more interdisciplinary, requiring additional data on the quantity and quality of surface and groundwater to plan solutions for water resource-related issues. GIS helps that all besides giving and analyze the data of remote sensing that are convenient and cheaper to collect when compared with field data collection.

Hydrological data can be classified in various ways. The most common classification includes temporal, spatial, and primary data categories [31]. Based on these criteria, data can be classified in a database and decision support system for planning and employing the most suitable strategies for water utilization, taking advantage of its intense data collection, storing, and analysis.

GIS will enable bridging the gap of acute data availability and give scenarios for policy and decision-making. Data distributed in time and space can be managed using GIS applications, and imitation systems can be developed using collected data. Water resources can be researched, analyzed, and efficiently utilized using GIS [32].

It is essential to develop, operate, and maintain the hydrological and meteorological data network. Regular reviews and redesign of the station network are necessary to ensure data quality and reduce errors. Having solid databases is a prerequisite of efficient management as it will enable establishing and fueling decision support systems "DSS" and information systems "ISs" that are significantly helpful in proper decision-making.

Decision support systems are commonly created and designed for various purposes in water resources management, such as involving data collection, creation of suitable scientific models, and determining decision-making mechanisms to achieve the system's goals. In the field of water management, these systems are often used to allocate water for specific purposes, such as agriculture or industry, considering factors such as quantity and quality in proportion to water availability [31].

However, a decision support purposed to efficient management and management of water resources could designed and operated [33]. A typical decision support system is designed based on the outputs of the data processing system, translating available information into practical and applicable actions [31]. This enables decision-makers to study various water management scenarios and choose the most appropriate one.

Decision support systems in water management commonly address tasks such as suitable decision-making in water allocation, analysis of appropriateness, and planning for the exploitation of water resources, development of water sources, study and monitoring of water quantity and quality, computation of water quantities and usage branches, and conservation of environmental and economic aspects of water resources. According to [31], these systems are needed and can be used for various purposes related to water resources in Sudan, the usages and applications include, not limited to:

1. Quantitative and qualitative assessment of the country's water resources.

2. Planning, designing, and operating water projects.

3. Environmental, economic, and social impact assessments for sound resource management and strategic planning.

4. Early prediction and safety measures for water-related risks.

5. Allocation of water among different users.

6. Determination of water quantities for industrial, domestic, and agricultural use.

7. Analysis of river characteristics and required uses of flowing water, such as fishing and transportation.

8. Environmental characteristics, such as damages to ecological systems.

In summary, GIS is a crucial technology and platform in order to connect the hydrological information systems to a database, along with information on water usage to make it ready for use in water resource planning and management. GIS is vital for Sudan, especially as an important tool for handling water resources planning to ensure that water is used properly and sustainably, in accordance with the predefined plans **Figure 6**.

5.2 IWRM approach

Integrated water resources management (IWRM) is a process that aims to coordinate the development and management of water, land, and their related resources in order to achieve maximum economic and social welfare in an equitable manner, while also preserving the sustainability of vital ecosystems [17].

The IWRM is a valuable approach for sustainably managing water resources. Its implementation is widespread across many countries worldwide and has proven effective in improving water security, reducing water pollution, safeguarding aquatic ecosystems, promoting economic growth, and enhancing human health [34].

Innovations in IWRM can contribute to improving the quality, sustainability, and efficiency of urban and agricultural water services by integrated and coordinated utilizing of water, in example, reusing household wastewater to meet the needs of both industrial and agricultural sectors. In Sudan, collaboration in shared water resources enhances and supports decisions toward water security sector. Adopting SWOT analysis is considered a valuable and systematic tool for strategic-level water management planning, formulating necessary operational policies, studying water, planning for its resource development, and optimal utilization and preservation (**Figure 7**) [9].

The philosophy of IWRM approaches is built upon a "global belief" and have an overarching perceptions on water and community and how they interact and

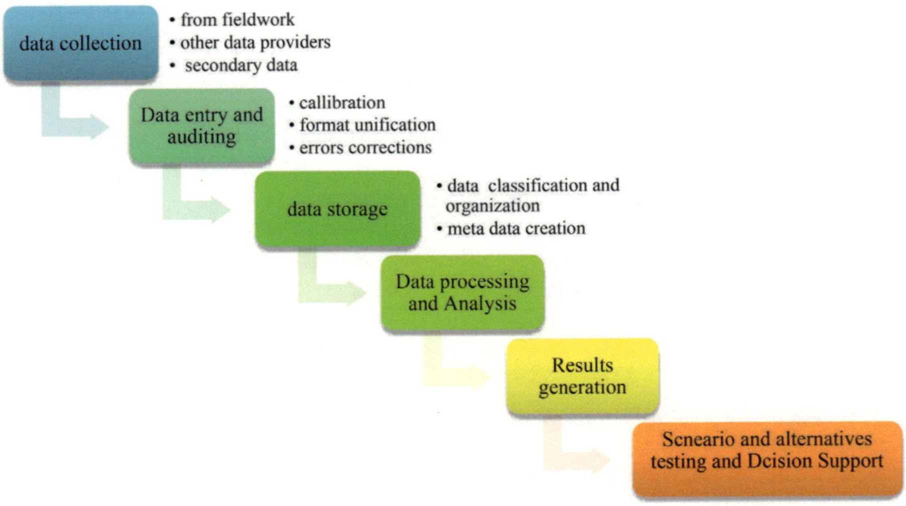

Figure 6.
Indicates how a typical DSS works. Source: Designed by the authors.

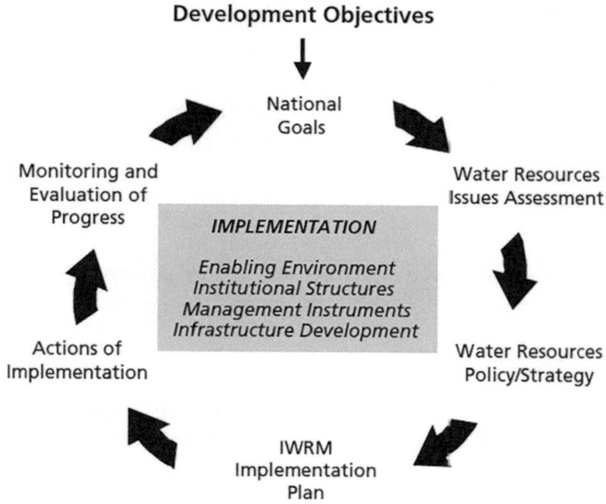

Figure 7.
IWRM planning and implementation stages. Source [34].

influence each other. For this, IWRM is theoretically based on the following principles: including, not limited to:

- Water is a finite and vulnerable resource.
- Water development and management should be based on a participatory approach that involves all stakeholders.
- Water should be managed in an integrated manner, considering its multiple uses and the interconnections between water, land, and other resources.
- Therefore, water management should be based on sound science and technology.
- Water should be managed in an equitable and sustainable manner, both within and between generations.

6. Suggested interventions and action plans for enhancement in Sudan

Sudan is facing environmental and ecological challenges, and the most important and threatening ones are related to weakened water resources and water scarcity such as drought and desertification. Because the greater part of Sudan population is rural communities of farmers and pastoralists, who are well-known for depending on rainwater for livelihoods.

The rural Sudanese are often displaced by changing landscapes and a lack of agricultural yield. As a result, demand for water rises in accordance to rising population and their different activities, while water supply remains limited and steady, if not decreased by pressing climatic factors. Several reasons contributed to this misery, to mention some, warming climate, population and urban growth, economic development, transboundary water issues, and water pollution.

To address these challenges and strengthen water sources, it is vital to undertake integrated techniques for water control. The integration of integrated water resources management (IWRM) processes and GIS techniques can improve decision-making and control strategies inside the face of changing hydrological and climatic patterns. GIS can be used as a spatial analytics tool for water tracking, vulnerability evaluation, and planning. The below listed interventions are suggested to improve water management practices in Sudan:

1. Conducting a comprehensive quantitative and qualitative assessment of the country's water resources. To enable planning, designing, and operating water projects.

2. Formulating and adhering to a national strategic plan that covers environmental, economic, and social impact assessments for water resource management.

3. Preparedness for future climate scenarios and anticipate their prospective role in the climate to enable establishing early prediction and warning and safety measures for water-related risks.

4. Planning for allocation of water among different users and sectors equitably in the light of development within the national strategic plan frame.

5. Well specifying of water quantities for industrial, domestic, and agricultural use. To enable detection, the deficit and follow the integrated water usage among various sectors.

6. Bridging the gap in knowledge and available datasets. Through unified data formats and integrated databases for all sectors and fields related to water resources.

7. Introducing a computerized geoinformation system, as a base for easy, affordable, quick, and (efficient and reliable) data collection and baseline surveys, which cover an in-depth study of the geography, environment, and resources of the country with quick and constant updates.

8. Finding a mechanism for international and regional coordination and cooperation to protect the countries interests that related to water.

9. Achieving the political and managerial stability and institutional reform.

7. Conclusion

Owing to its inherited physical and human geography, Sudan is facing dire ecological challenges, such as water scarcity and desertification. Because the greater part of Sudan population is rural communities of farmers and pastoralists, whom are well-known for depending on rainwater for livelihoods. The rural Sudanese are often displaced by changing landscapes and a lack of agricultural yield. As a result, demand for water rises, while supply remains limited and steady, if not decreased by pressing climatic factors.

Water stands as a principal factor for development and crossing from pastoral community toward the industrial society. Therefore, the development and prosperity of the nations are strongly linked with providing development projects for rural communities and enhancing living standards. This transformation, from pastoral to agricultural, and subsequently to industrial society, forms the basis for the stability of these communities. Thus, emphasis on securing water resources plays a crucial role in facilitating significant developmental projects, such as irrigation and animal production initiatives.

Sudan, as any of arid environments, is highly susceptible to degradation, and water resources must be protected from waste, pollution, or unsustainable use. Therefore, it is necessary to reduce the adverse hydrological and environmental impacts of water consumption, as well as consider how internationally shared persistent water resources could be used to enhance scarce water resources. Twining GIS and IWRM together would significantly help to address the challenges hampering of water sources to reach the state of security and sustainability.

Acknowledgements

With immense gratitude, the authors would like to extend thanks and acknowledgment to their former supervisor, the associate professor at University of Khartoum Dr. Badr Eldin Taha Osman for his dedication, constructive guidance, fatherly advice and support, and continuous motivation for scientific research.

Author details

Abdelbagi Y.F. Adam[1,2*] and Mohamed B.O. Osman[1,2*]

1 Institute of Geomatics and Civil Engineering, University of Sopron, Sopron, Hungary

2 Department of GIS and Cartography, University of Khartoum, Khartoum, Sudan

*Address all correspondence to: abdoyanes2016@gmail.com and mohamedbabekir95@gmail.com

References

[1] du Plessis A. Water resources from a global perspective. In: South Africa's Water Predicament. Water Science and Technology Library. Vol. 101. Cham: Springer; 2023. DOI: 10.1007/978-3-031-24019-5_1

[2] Bekele DA, Bona SK, Haji HS. Water policy for sustainable management: A review. International Journal of Applied Agricultural Sciences. 2021;7(3):110-118. DOI: 10.11648/J.IJAAS.20210703.11

[3] Lunegova IV, Tyts VV, Lopatin SA. Water and its importance for human life and health. Формулы фармации. 2023;**4**(3):50-57. DOI: 10.17816/phf115210

[4] Kryvega L, Sukhareva K. Water as a factor of social development and contamination of countries regarding water resources. Humanities Studies. 2022;**13**(90). ISSN 2708-0390 (Print) 2708-0404 [Online]. DOI: 10.26661/hst-2022-13-90-04

[5] The Social Life of Water. 2022. DOI: 10.1515/978085745967 [Assessed: January 31, 2024]

[6] Adamopoulou J, Frantzana A, Adamopoulos IP. Addressing water resource management challenges in the context of climate change and human influence. European Journal of Sustainable Development Research. 2023;7(3):em0223. DOI: 10.29333/ejosdr/13297

[7] Balata EE, Pinto H, Moreira da Silva M. Latent dimensions between water use and socio-economic development: A global exploratory statistical analysis. Regional Sustainability. 2022;**3**(3):269-280. DOI: 10.1016/j.regsus.2022.09.004

[8] Hosseiny SH, Bozorg-Haddad O, Bocchiola D. Water, culture, civilization, and history. In: Economical, Political, and Social Issues in Water Resources. 2021. pp. 189-216. DOI: 10.1016/B978-0-323-90567-1.00010-3

[9] Hyaty AEE. National Water Security: From Planning to Practice. 2021. Available from: https://www.researchgate.net/publication/357450312_alamn_almayy_alqwmy_mn_altkhtyt_aly_altnfydh_National_Water_Security_From_Planning_to_Practice

[10] Janardhanan R. Water Management: A Key to Sustainable Development. In: Basheer AK, et al., editors. Handbook of Research on Future Opportunities for Technology Management Education. Hershey, Pennsylvania: IGI Global; 2021. pp. 387-400. DOI: 10.4018/978-1-7998-8327-2.CH023

[11] Abdo GM, Salih A. Challenges facing groundwater management in Sudan [Internet]. 2012. Available from: https://garj.org/garjpas/11/2012/1/1/challenges-facing-groundwater-management-in-sudan [Assessed: Jun 1, 2024]

[12] Encyclopedia Britannica. 2024. Available from: https://www.britannica.com/place/sudan [Accessed: Dec 19, 23]

[13] REDD. Reducing emission from deforestation and forest degradation readiness program, land use in Sudan study report. 2017

[14] Hamadalnel M, Zhu Z, Gaber A, Iyakaremye V, Ayugi B. Possible Changes in Sudan's Future Precipitation under the High and Medium Emission Scenarios Based on Bias Adjusted GCMs. Atmospheric Research. Vol. 269. Amsterdam, Netherlands: Elsevier;

2022. p. 106036. DOI: 10.1016/j.atmosres.2022.106036

[15] Grey D, Sadoff CW. Sink or swim? Water security for growth and development. Water Policy. 2007;**9**(6):545-571

[16] UN University. Water Security & the Global Water Agenda: A UN-Water Analytical Brief. Canada: Ontario; 2013

[17] Mohamed IAW. Economics of water use in semi-arid environment: Darfur case study. Emerging Markets Economics, Environmental and Social Aspects Journal. 15 Mar 2011;**2**(24). Available from: https://ssrn.com/abstract=1769630

[18] Fanack Water, (a website). Water resources in Sudan. Available from: https://water.fanack.com/sudan/water-resources-in-sudan/ [Accessed: Dec 26, 2023]

[19] Proceedings of the second expert consultation on national water policy reform in the near east. Available from: https://www.fao.org/3/ad456e/ad456e0f.htm#TopOfPage [Accessed: Dec 21, 2023]

[20] ESCWA. Moving towards water security in the Arab region. United Nations Economic and Social Commission for Western Asia. 10 September 2019. Available from: https://archive.unescwa.org/publications/moving-towards-achieving-water-security-arab-region

[21] Hamadalnel M, Zhu Z, Gaber A, Iyakaremye V, Ayugi B. Possible changes in Sudan's future precipitation under the high and medium emission scenarios based on bias adjusted GCMs. Atmospheric Research. 2022;**269**:106036

[22] UNEP - The UN Environmental Program. Sudan first state of environment and outlook report 2020. 2020. Available from: http://www.unep.org/resources/report/sudan-first-state-environment-outlook-report-2020

[23] Heyns P. Water conservation in the arid and semi-arid regions. In: Hoekstra AY, Savenije HHG, editors. Water Resources Management. EOLSS; 2009;**1**:113-150. ISBN: 978-1-84826-177-8 (eBook)

[24] WB. World Bank, Sudan—Climatology — Climate change knowledge portal. 2024. Available from: https://climateknowledgeportal.worldbank.org/country/sudan/climate-data-historical [Accessed: January 27, 2024]

[25] International Decade for Action "Water for Life" 2005-2015. Focus areas: Integrated water resources management (IWRM). Available from: https://www.un.org/waterforlifedecade/ [Accessed: Dec 21, 2023]

[26] Khalifa M, Nadir AE, Ribbe L, Schneider K. Spatio-temporal variations in climate, primary productivity and efficiency of water and carbon use of the land cover types in Sudan and Ethiopia. Science of the Total Environment. 2018;**624**:790-806. DOI: 10.1016/j.scitotenv.2017.12.090

[27] Nicholson SE. Climate and climatic variability of rainfall over eastern Africa. Reviews of Geophysics. 2017;**55**(3):590-635. DOI: 10.1002/2016rg000544 [Accessed: Dec 19, 2023]

[28] Hawchar L, Naughton O, Nolan P, Stewart MG, Ryan PC. A GIS-based framework for high-level climate change risk assessment of critical infrastructure. Climate Risk Management. 2020;**29**:100235. DOI: 10.1016/j.crm.2020.100235

[29] Benefits of GIS data design in water resource management and optimization.

Utilities one n.d. Available from: https://utilitiesone.com/benefits-of-gis-data-design-in-water-resource-management-and-optimization [Accessed: Dec 21, 2023]

[30] Khatami S, Khazaei B. Benefits of GIS application in hydrological modeling: A brief summary. Vatten: tidskrift för vattenvård /Journal of Water Management and research. 2014;**70**(1):41-49

[31] Hyaty AEE. A guide to collect water data and its measuring methods. 2022. Available from: https://www.researchgate.net/publication/363924359

[32] Zhang Z, He Y, Chen X, Tan Q. Improvement of WEAP model considering regional and industrial water distribution priority and its application. Journal of Hydrology: Regional Studies. 2023;**47**:101414. DOI: 10.1016/j.ejrh.2023.101414

[33] Raza A, et al. Water resources and irrigation management using gis and remote sensing techniques: Case of multan district (Pakistan). In: Pande CB, Kumar M, Kushwaha NL, editors. Surface and Groundwater Resources Development and Management in Semi-arid Region. Springer Hydrogeology. Cham, Switzerland AG: Springer; 2023. pp. 137-156. DOI: 10.1007/978-3-031-29394-8_8

[34] UNEP-UN Environment Program. What is the integrated water resources management?. 2017. Available from: http://www.unep.org/explore-topics/disasters-conflicts/where-we-work/sudan/what-integrated-water-resources-management

Chapter 6

Water Resource Management Using Geographical Information System and Remote Sensing Technologies: A Case of Jazan, Saudi Arabia

Saeed Q Al-Khalidi Al-Maliki, Ahmed Abdelmotlab Ahmed, Mohammed Ahmed Al-Ghobiri and Osman Ali Nasr

Abstract

This study identifies the importance and ability of geographic information systems and remote sensing technologies to manage and monitor surface water in the Jazan region of Saudi Arabia. This study uses several methods to establish a geographical database for the study area: hydrologic and water quality data, a digital Geographical Information System (GIS) elevation model, the spatial distribution and investigation of water samples in Jazan and watershed delineation and flowing tracing tools. The proposed spatial data architecture attempts to implement flood prevention tools and practical land-use proposals, as well as the design of a computerised GIS database for flood risk assessment and mapping.

Keywords: GIS, Saudi Arabia, spatial data, water resources, remote sensing

1. Introduction

In the last half-century, the Earth has faced drastic and spectacular changes in its tidal and planetary movement, as identified by the appearance of natural disasters in the equatorial zones. Governments must address the natural destruction faced by economically growing cities. Scarce planning in the agricultural field, which is controlled by conventional water authorities, has raised the problem of extra development needed in underground water. This revolution has benefited humankind by reducing malnourishment in developing countries and providing drinking water to rural and urban areas, while the lack of prevention in flood water flow and supply of drinking water has had negative ecological impacts on the geographical condition of the land [1].

Water resources play an essential role in providing freshwater for irrigation, industrial activities, and drinking, but extracting the maximum level of groundwater from alluvial aquifers makes them vulnerable to pollution-related activities. This effect has been widely studied [2–5]. High temperatures, high underground evaporation rates and minimal rainfall define the Kingdom of Saudi Arabia (KSA) as one of the world's driest countries. Additionally, agricultural advancements and industrial growth play an essential role in forming unmanageable groundwater concepts and mitigating water quality [6].

Many areas of Saudi Arabia consist of hilly, vegetation-free slopes and steep, mountainous terrain. Freshwater management is crucial for this area. The sub-substantial condition in many areas of the KSA has reflected the increased rate of Earth flooding [7–14].

Water management resource issues are resolved by improving the water supply, non-point source pollination and groundwater-surface quality. Several of these sources require understanding the fundamental features of physical, biological, social, and economic components and better information regarding operating these components together within the watershed [15]. Water management resources are linked to many watershed components, such as rivers, wetlands, and groundwater. This system requires methods to evaluate land use and watershed management operational practises, and watershed simulation models are required to reduce uncertainty in decision making.

According to the study, to interrelate the information of a specific global area, remote sensing technologies provide valuable updates to the spatial information on natural resources and the climatic and geographical conditions of the remote area. GIS integrates and analyses multi-layered spatial information and has a variety of specific formats for effective planning for watershed development, along with Integrated Sustainable Development guidelines and decision rules. Linear Imaging Self scanner (LISS-III) satellites are advanced in collecting data based on land cover hydrogeomorphology, which is essential for water resource planning and development. On the basis of environmental conditions, GIS analyses spatial data integration [16].

2. The problem

The Jazan region, like the rest of the regions of the Kingdom of Saudi Arabia, suffers from the scarcity of available groundwater resources as a result of its geographical location, and a portion of the surface water is lost through a lot of valleys without optimal exploitation, due to the region's topography and its steep slopes. Therefore, this study sheds light on the possibility of benefiting from surface water and contributing to alleviating the severity of the water crisis in the study area. The problem of the study can be summarised in the following points:

- Failure to make maximum use of rainwater in the study area.
- The phenomenon of flooding and its negative effects on the infrastructure in the study area.
- Late and sudden reporting of flood phenomena.
- Traditional mechanisms used to deal with these phenomena.

In addition, after reviewing previous studies that related to this study subject, the researchers believe that there are technical additions that distinguished the study, including:

1. Using cloud storage technologies, as well as relying on the NoSQL structure, suits such studies, which are characterised by the abundance of data generated from the sensors used in the study, which provides huge data that can be used in future studies.

2. The study provides smart application interfaces for dashboards that will help in making decisions.

3. Integrating some machine learning algorithms to predict the values of phenomena in different places in the study area.

4. The technologies used are free, and the application is available to everyone.

Moreover, the researchers emphasise that this study is a unique contribution to the area of the Jazan region as it has a lot of valleys and very heavy rainfall, especially during summer months.

Therefore, this study can be significant research that can be added to the previous related studies in the application of GIS.

3. Objectives

This study seeks to achieve some objectives, including:

- Determine the locations of depressions or basins to retain water and use it in agriculture and other areas.

- Reducing the negative effects of flooding in the study area.

- Providing accurate information to manage, control, and reduce the severity of the flood phenomenon.

- Using modern technologies to deal with the flood phenomenon.

- Build a cloud database for the study area.

- Reducing wastage of rainwater.

4. Study area

The Jazan region is located in the southwestern part (East: 42–43.8 and North: 16.5–17.0) of Saudi Arabia, consisting of more than 100 islands. Its area is about 13,500 km^2. The landform of Jazan is mostly alluvial, with a temperature of about 28C, relative humidity of 62% and annual precipitation of 62 mm. It consists of many valleys and drainage channels directed towards the Red Sea, as shown in **Figures 1** and **2**.

It is situated on the coast of the Red Sea, serves as the region's most important large agricultural land and is characterised by different environmental and climatic changes. This region consists of several governmental and administrative centres [19].

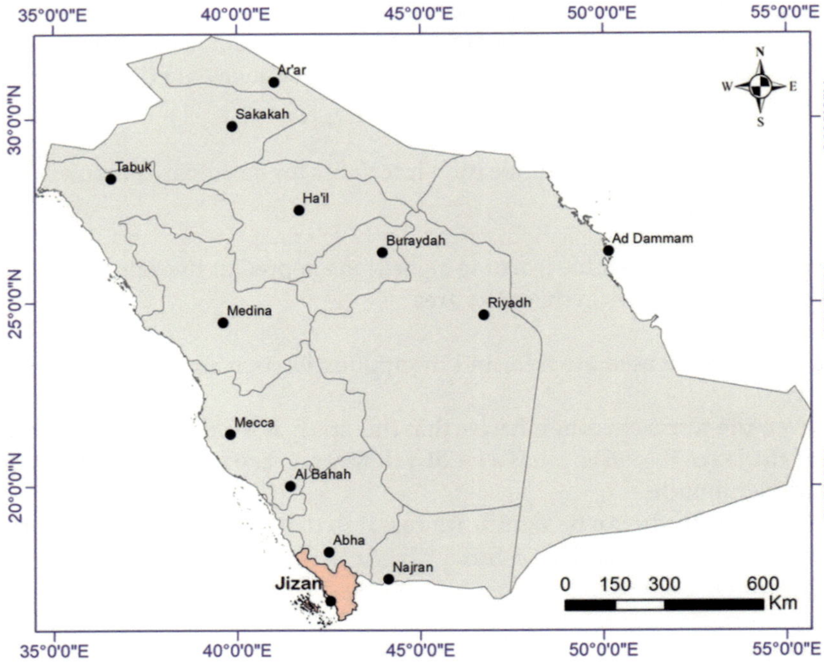

Figure 1.
Kingdom of Saudi Arabia. Source: [17].

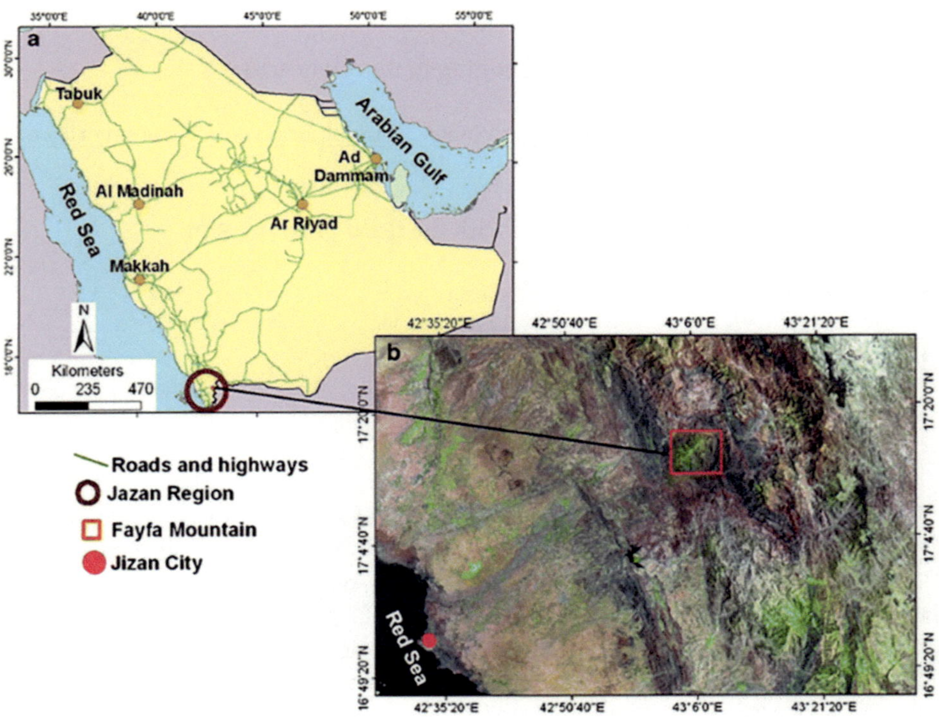

Figure 2.
GIS map of Jazan region, Saudi Arabia [18].

The economic importance of this Saudi Arabian city lies in its ideal location at the mouth of the Red Sea, where it serves the markets of Europe, Asia and East Africa by transporting raw materials to serve the manufacturing facilities of the Jazan [20]. Rainfall ranges from 100 to 450 millimetres, depending on the elevation of the site, and these rains are often dense and heavy, generating massive and sudden floods due to the rainfall in mountainous areas and steep slopes towards the coast, where the central valleys of the region often run from east to west towards the Red Sea coast. The Jazan region's climate is highly affected by tropical winds. The coastal plain climate is moderate in winter and hot and humid in summer, resulting in rainfall. Due to Jazan's environment and seacoast, it is Saudi Arabia's naval industry hub, and the government also wants to convert it into an agricultural hub due to its rich underground water reserves.

The KSA is highly affected by arbitrary climatic events that cause considerable damage to agricultural zones, railroads, and highways [21]. According to the study, about 34 Million kilometers3 (Mkm3) of water is present on Earth, and 22% of the groundwater accounts for 97% of the potential potable water for human consumption [22], as shown in **Figure 3**.

In arid to semiarid regions where there is low rainfall and limited freshwater resources, there is greater emphasis on the use of groundwater as a source of freshwater. The freshwater acquaintance has an accessible depth, therefore giving rise to salinity, seawater ingression, and water pollution due to over-drafting, which degrades the quality of soil [22–25].

Figure 3.
Flood risks on Jizan-Abha highway [20] (a) showing the flooding of the Jazan-Abha road on 15/4/1440 H, (b) showing the flooding of the sides of the road. Picture (c) shows the floodwaters passing under the Jazan-Abha road as a result of the Wadi Bayad floods. Picture (d) shows the Jizan road-Abha collapsing [20].

5. Materials and methods

Geologists and researchers consider remote sensing (RS) and GIS as techniques for evaluating flood hazard impacts and agricultural infrastructural change in mountainous areas [26–29]. The water resource system requires the assessment and management of geographical activities, requiring the handling of multiple forms of spatial data.

The study uses a systems methodology in designing databases and building the proposed application, as well as a spatial analysis methodology through the following steps:

- Obtain a recent satellite image of the area in which the study area is located.
- Determine the study area accurately from approved sources of maps and reference them geographically to suit the method of the coordinate system used in satellite images (geographical coordinating system).
- Obtain a map of the elevations covering the study area.
- Apply mosaic or clip technology to satellite images to cover the study area.
- Conduct Hydrologic analysis of the map using ArcGIS Pro.
- Determine the proposed locations for the bubble sensors and connect them within a wireless network.
- Design a NoSQL database on Google Firebase to store sensor data.
- Design a mobile application interface and display and analyse sensor data.

6. Tools and techniques used

Earth explorer: The tool provides users with the ability to query, search, and request satellite images, aerial photos, and mapping products from several sources.

1. ArcGIS Pro: Geographic Information Systems technology, through which the following tools were used:
2. Georeferencing tool: for returning and geometric correction of satellite images.
3. Geoprocessing tool: To process spatial data.
4. Hydrologic tool: Tools are used to model the flow of water across the surface.
5. Firebase: Google's web and mobile app development platform that helps developers create apps and games that users will love.
6. Flutter: It is an open-source framework from Google for creating beautiful, locally compiled, cross-platform applications from a single code base.

7. Data source

It consists of a set of thematic maps, such as topographic and level maps. I also used aerial photographs for the year 2022 with a resolution of 15 cm and contour lines with a vertical interval (5 m), and I used the shuttle radar topography mission (SRTM) satellite images from the Earth Explorer website of the study area dated October 16, 2023, a pixel size of 30×30.

8. Major contributions and significance of GIS

GIS offers the most powerful tool for studying the various combinations of locations and enhancing the performance of the simulation models to improve area knowledge. Stimulation models provide decision-making geologists with interactive tools for understanding the physical stem and judging how the installed support system affects the overall system [15].

The most important role is in the development of distributed hydrologic models and an improved understanding of the spatial aspects of the distribution and movement of water in landscapes. The outcomes of administrative policies and management strategies for small plots are mainly focused on watersheds and jurisdictions.

9. Data management of GIS distribution

Conversion of such databases into digital form can be accomplished for the future investigation of water distribution, sample collection and utilisation, and for preparing location-specific land and water resource development plans in the GIS environment by implementing such databases with socio-economic data.

10. Determination of hydrologic data and water quality data

Data collection for a water resource management spatial data model is based on a 1:25,000 scale linking drainage, watershed, transport network, surface water bodies, hydrogeomorphology, etc., via GIS and RS technologies. This data includes terrestrial ecosystem, watershed, and river network hydrologic modelling studies.

11. Digital GIS elevation model

The impact of water resource management in GIS is simulated by stimulating the development of source pollution models and the analysis of the predicted movement of water-related contaminants in natural and anthropogenic landscapes. Radar technology is also used to create the potential for hydrologic studies at possible regional scales.

12. Spatial interpolation tools

The data are often distributed with hydrologic and water quality data by using new satellite sensors that work at a finer spatial and temporal resolution to regional climate stations.

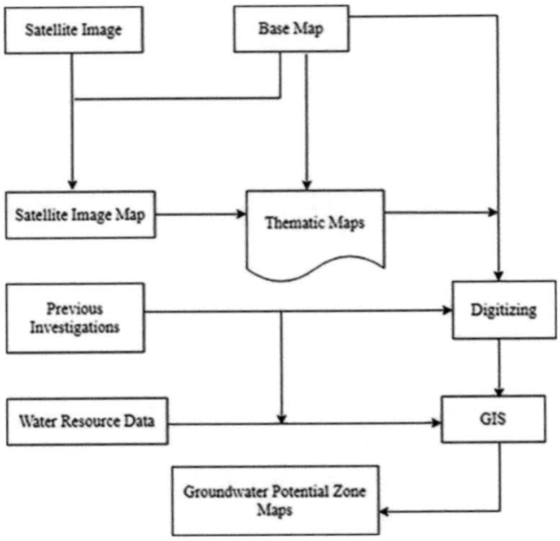

Figure 4.
RS and GIS mapping.

13. Potential zone mapping of RS and GIS

Improved prediction and accuracy of promising locations and their groundwater potential can be gained by identifying pertinent geomorphological and geological metrics, allocating adequate weights, preparing subsequent thematic maps, and using high-quality remotely sensed data. RS- and GIS-based frameworks are shown in **Figure 4** for groundwater potential zone mapping.

14. Spatial distribution and investigation of water sample in Jazan

Spatial distribution and the identification of groundwater have different objectives in the Jazan region. Extraction and investigation of water samples from different regions of Jazan were initiated using a modified GIS system. The modified system's musing geostatistics (kriging) and splines create spatial and spatiotemporal models of land surfaces, climatic phenomena like precipitation and temperature, soil properties, and water quality from measured data. This method was modified according to the regional requirement for proper investigation: (a) the water component layer in the aquifer meets drinking standards; (b) new GIS layers must identify appropriate locations for drinking water in the groundwater aquifer; and (c) significant water components are produced by different component combinations [30]. There are four basic formats for representing and storing data: satellite images, vector maps, aerial maps, and GPS data. These represent cartographic features, much like maps or scaled drawings, with scattered points and lines. Each type of mapping model represents a separate feature class. All of these formatted image files are presented as layers in a GIS database. The information was collected from various sources that facilitate the analysis and formulation of the proposed water resources management model, including agriculture and environmental aspects, to assist in decision-making and developing plans (**Figure 5**).

The GIS database system comprises data models that identify the spatial relationships between multiple datasets. They involve an input dataset and a tool for analysing and generating new information, resulting in an output dataset.

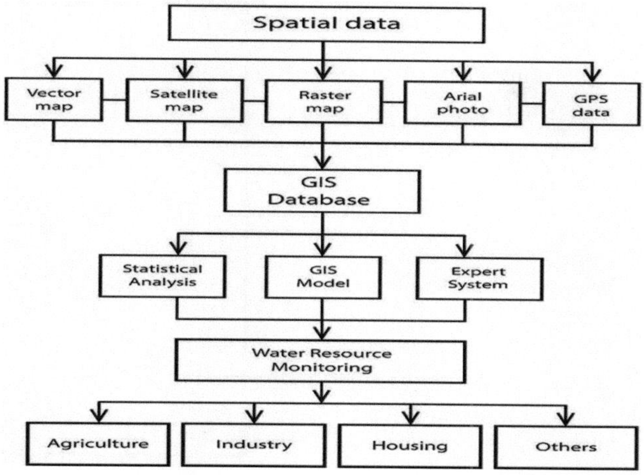

Figure 5.
Spatial data for water resources monitoring architecture.

15. Water shed delineation and flowing tracing tools

Hydrologic modelling and water resource management require algorithms that support portioning hydrologic units and provide the basic parameters required for flow routing and hydrologic models. Identification of potential sources of non-point source pollution in flow tracing has allowed the simulation of the movement of water, sediment, and other pollutants through landscapes [31] by developing new tools for mapping algebra-supporting computation with spatiotemporal data.

16. Visualisation tools' impacts on computer cartography

GIS communication of water resource data increases the availability of maps generated efficiently by cartographic tools to visualise the development and present the model in the mountainous or tropical region of Jazan [32]. Many studies are presented to illustrate models of water movement and identify saline and drinking water extraction by modified processes and patterns. (**Figure 6**).

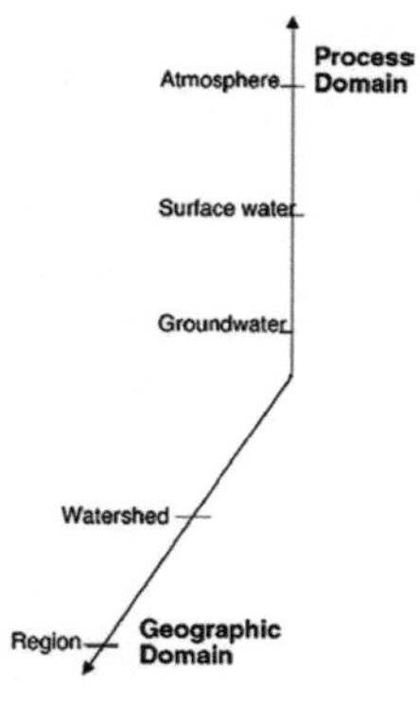

Figure 6.
Mapping by visualisation tool.

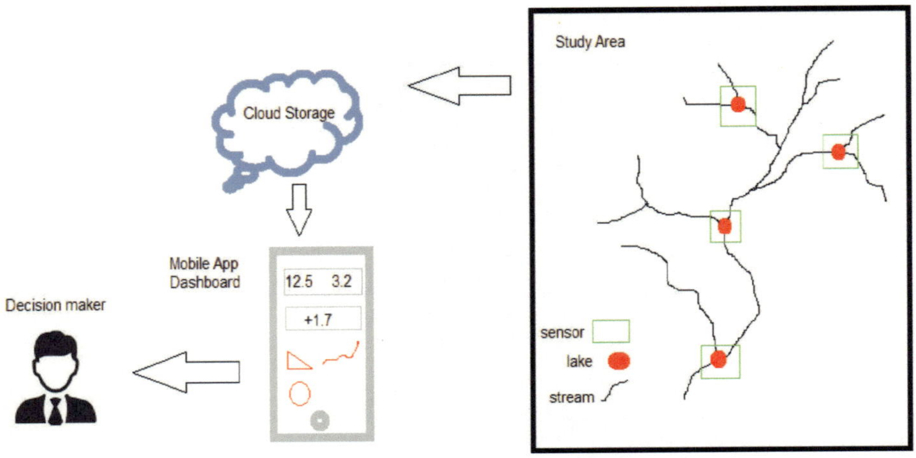

Figure 7.
System proposal workflow.

17. System workflow

- The sensors read the level of water in the lakes and send it to the server.
- The server stores the data and sends a copy to the mobile application.
- The Oracle App processes data and extracts reports to the mobile application.
- The decision-maker takes readings of the statistics and possible predictions and their percentages.
- The decision-maker issues appropriate decisions (**Figure 7**)

18. Data processing

The obtained datasets have different formats and scales, so to facilitate the precise processing and interaction between them, a GIS database using ArcGIS 10.2 software was developed. All datasets were using the universal transverse mercator (UTM) coordinate system, zone 37. Digital elevation models and topographic maps were used to extract all streams and catchment areas from the images. Interpretation of satellite images was carried out to determine the areas of urban expansion, and HEC-RAS was used to determine the inundation zones. The dark subtraction method was used to reduce the atmospheric effect on the reflectance values using Landsat data imported into ENVI 5.3 software [33]. The stratified random sampling method was then used to assign a proportional number of regions of interest to each land-use class based on its area and homogeneity.

19. Results and discussion

This paper relies on free satellite images (Landsat 8) at the planning stage and can obtain high-resolution images for the implementation stage. **Figure 8a** shows the part of the study area taken from Landsat 8.

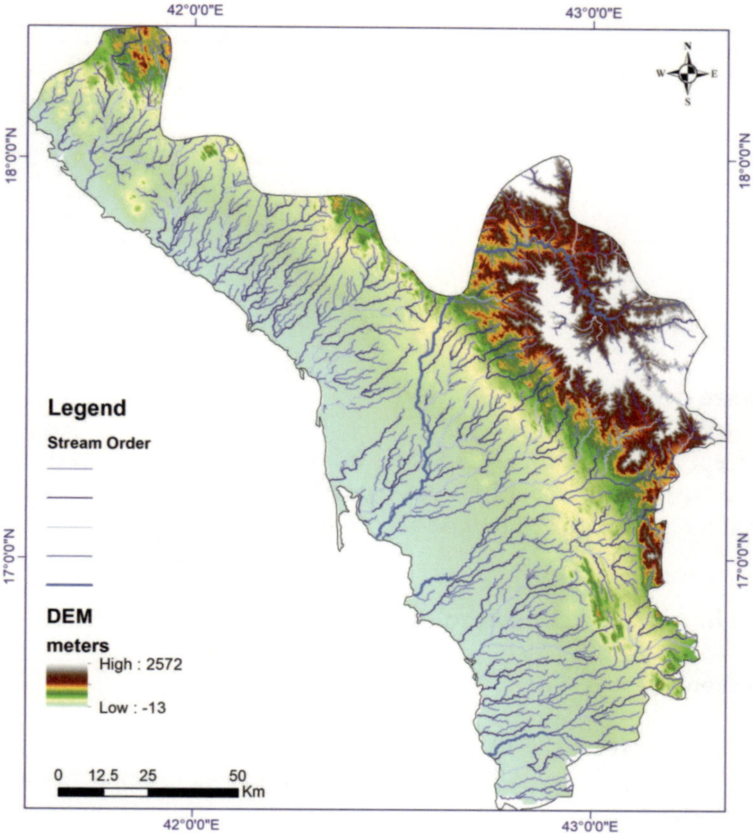

Figure 8.
Base maps.

A set of statistical and spatial procedures was carried out to obtain **Figure 8b** and get the watersheds in the study area:

- Hydrology toolset
- Flow direction
- Flow accumulation
- Flow length
- Stream order
- Position of Watershed

Based on satellite image processing outputs and this study, the layers below were agreed upon, and a geographic database was designed using RS and GIS technologies. These parameters were identified to obtain the appropriate estimation in the software of the regional water resourcing management (**Figure 9**).

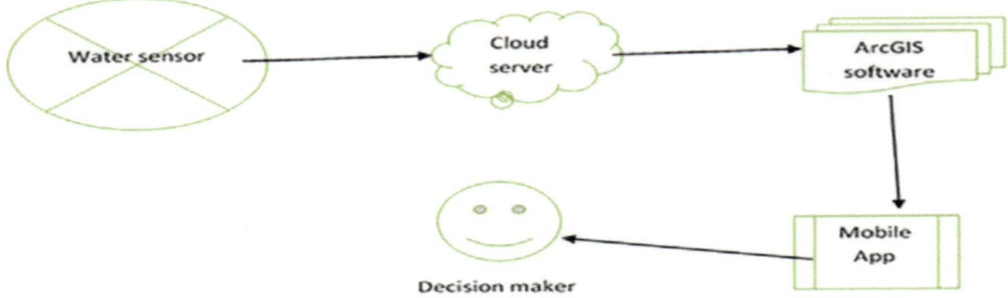

Figure 9.
Database design.

20. Database design

The concept of non-relational databases was relied upon to suit these projects to store and manage the huge data recorded from the readings of the sensors used, and the author also depend on the Firebase-database service.

21. Project hardware and technologies

21.1 Water level sensors

The water sensor reads the water level in the tank and sends the reading over the internet to the proposed server. The water level is displayed in a smartphone app as a vertical bar illustrating the percentage capacity of the water. The app sounds an alarm when the water level reaches a specific value.

21.2 Cloud storage

Firebase will be used for the project database for extracting the water sensors, and cloud storage will be used for the big data.

21.3 ArcGIS and Oracle applications

The Oracle database management system is needed to manipulate the descriptive data (sensors), and ArcGIS software is used to design the necessary layers. **Figure 10** shows some of the functions that are included in the project.

21.4 Android app

The mobile app will be designed to read sensor data and give the decision-maker statistics and predictions. The Ministry of Environment, Water and Agriculture in Saudi Arabia aims to conserve water sources and promote rainfall and surface water to ensure the provision and distribution of water. This can be achieved through specific strategies such as raising public water management standards, managing water resources, rehabilitating water collection and meteorological data collection networks, building multi-purpose dams, and creating water facilities to support growing populations, new towns and villages, industrial cities and agricultural land reclamation.

DOI: http://dx.doi.org/10.5772/intechopen.1004352

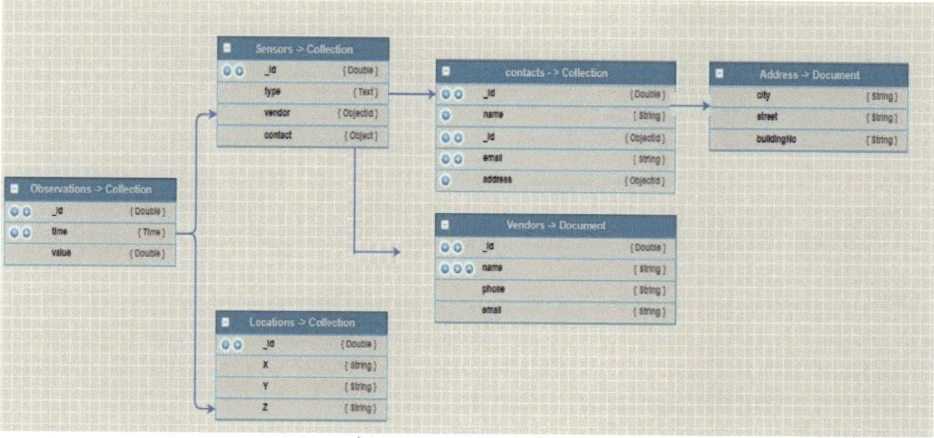

Figure 10.
Some of the proposed system functions.

In addition to hydrological applications, GIS can study rainfall flow systems in the Jazan region and capture watersheds, which will significantly help identify the best distribution of flood wells for the population in these areas. In this way, the best areas for irrigation can be determined, thereby promoting agricultural production and access to water. The GIS approach is also used for water network planning, engineering, operations, maintenance, financing, and management functions. Dam construction requires determining the dam's location and the lake's size. GIS can identify the best water conservation practices and address where potential risks may arise.

This study aims to create a database of the morphometric and hydrological characteristics of dry valleys within the Jazan region in Saudi Arabia. Likewise, previous studies have discovered the existence of groundwater in acquirers in Hada Al-Sham [33, 34]. The study by Madani [35] found the problem of probable groundwater demarcation using RS and GIS in the Wadi Yalamlam and Makkah Jurisdiction catchment areas in west KSA. The study by Swain [36] used soil and water assessment tool (SWAT)-based hydrological modelling. Using the stream flow values inside a basin, the hydroelectric potential of that basin is calculated. Therefore, it is crucial to estimate basin stream flow correctly when designing hydroelectric structures. In other words, the runoff values dictate the size and, thus, the cost of these structures (**Table 1**).

#	Technology	Previous studies and projects	The current study
1	Application interface	Website	Mobile app
2	Database structure	SQL	NoSQL
3	IoT (Internet of Things)	Not available	Available
4	Availability of information	Not available	Available
5	Users	Decision makers	For everyone
6	Artificial intelligence	Not available	Machine learning algorithms
7	The information	Dashboards	Text reports

Table 1.
A comparison between the tools that have been used in similar studies and projects and the current study results to date.

22. Recommendations

After studying the surface water of the Jazan region, several results were carefully discussed, and it was found that GIS and geospatial tools were needed. In addition, this research provides recommendations that could help decision makers in the Ministry of Environment, Water, and Agriculture and its branch in Jazan understand the current situation regarding heavy rainwater and valleys. These recommendations could also help identify processes that would benefit from applying GIS and geospatial tools.

The researchers have visited the dams in Jazan and the multiple valleys that flow to the sea. There is a clear need for GIS and geospatial tools to manage and regulate the flow of rainwater in these valleys. This regulation will take advantage of the rainfall and help to prevent damage to farms and houses.

The agricultural sector in the kingdom uses mostly non-renewable groundwater, with some of its demands met by surface water and renewable groundwater. However, the Saudi Government has introduced some measures to partially reduce excessive pumping from groundwater reserves. The proposed spatial data water management strategy is the most crucial development in this context. The most significant steps include screening the availability of ground resources and the needs of the public and private sectors, assessing the environment, identifying resource areas, data management, decision making, and resource auditing.

The availability of groundwater is a significant issue in Saudi Arabia, and the country faces prolonged problems with drinking water scarcity. To reach self-sufficient levels of water intake, technical and non-technical measures must be linked, including raising public awareness. Based on the proposed architecture, there is a likelihood of a low level of public awareness about the water shortage problem in Saudi Arabia and a lack of measures taken by local water authorities in the Jazan region to address this issue. Extensive work is urgently needed to increase water awareness through media campaigns that explain the importance of water resources.

The findings revealed that urbanisation is the critical factor influencing agricultural regions. Seasonal irrigation water requirements can be handled both within a particular region and over a large area. To optimise water management in irrigation networks, irrigation water supply should be aligned with irrigation water needs. Water distribution between old and cultivated lands should be controlled.

In most cases, the failure of watershed development is significant because adjustments to numerous natural facets result in a lack of natural resource awareness. Scientific surveys should be conducted to generate optimum use of current natural resources such as water, vegetation, and land in the watershed. This will be further used to monitor watershed development in the future.

This research idea is considered a preliminary investigation proposal. Further work is needed to describe water awareness campaigns and offer financial support for implementing the spatial data model structure. Therefore, these researchers support programmes addressing water resources in the Jazan region that use GIS to manage the development of water sources and benefit from modern irrigation technologies.

DOI: http://dx.doi.org/10.5772/intechopen.1004352

Author details

Saeed Q. Al-Khalidi Al-Maliki*, Ahmed Abdelmotlab Ahmed,
Mohammed Ahmed Al-Ghobiri and Osman Ali Nasr
MIS Department, College of Business, King Khalid University, Abha, Saudi Arabia

*Address all correspondence to: salkhalidi@kku.edu.sa

References

[1] Ramirez-Pulido J, González-Ruiz N, Gardner A, Arroyo-Cabrales J. A list of recent land mammals of Mexico. USA: Special Publications of the Museum of Texas Tech University; 2014;**63**:1-69

[2] Su C, Zhu Y, Abbas Z, Huq M. Sources and controls for elevated arsenic concentrations in groundwater of Datong Basin, northern China. Environmental Earth Sciences. 2016;**75**(7):1-13. DOI: 10.1007/s12665-016-5359-1

[3] Huq ME, Su C, Li J, Sarven MS. Arsenic enrichment and mobilization in the Holocene alluvial aquifers of Prayagpur of southwestern Bangladesh. International Bio Deterioration and Biodegradation. 2018b;**128**:186-194. DOI: 10.1016/j.ibiod.2018.01.008

[4] Huq M, Su C, Fahad S, Li J, Sarven M, Liu R. Distribution and hydro geochemical behavior of arsenic enriched groundwater in the sedimentary aquifer comparison between Datong Basin (China) and Kushtia District (Bangladesh). Environmental Science and Pollution Research. 2018a;**25**(16):15830-15843. DOI: 10.1201/9781351046633-50

[5] Huq ME, Fahad S, Shao Z, Sarven MS, Al-Huqail AA, Siddiqui MH, et al. High arsenic contamination and presence of other trace metals in drinking water of Kushtia district, Bangladesh. Journal of Environmental Management. 2019;**242**:199-209. DOI: 10.1016/j.jenvman.2019.04.086

[6] Ahmed I, Nazzal Y, Zaidi FK, Al-Arifi NS, Ghrefat H, Naeem M. Hydrogeological vulnerability and pollution risk mapping of the Saq and overlying aquifers using the DRASTIC model and GIS techniques, NW Saudi Arabia. Environmental Earth Sciences. 2015;**74**(2):1303-1318. DOI: 10.1007/s12665-015-4120-5

[7] Coumou D, Rahmstorf S. A decade of weather extremes. Nature Climate Change. 2012;**2**(7):491-496. DOI: 10.1038/nclimate1452

[8] Mambretti S, De Wrachien D, Sole A. Flood-risk assessment and hazard mitigation measures: Case studies and lessons learnt in Italy. WIT Transactions on Ecology and the Environment. 2008;**118**:17-26. DOI: 10.2495/friar080021

[9] Avinash S. Flood related disasters: Concerned to urban flooding in Bangalore, India. International Journal of Research in Engineering and Technology. 2014;**03**(28):76-83. DOI: 10.15623/ijret.2014.0328013

[10] Bathrellos GD, Karymbalis E, Skilodimou HD, Gaki-Papanastassiou K, Baltas EA. Urban flood hazard assessment in the basin of Athens Metropolitan City, Greece. Environmental Earth Sciences. 2016;**75**(4):1-14. DOI: 10.1007/s12665-015-5157-1

[11] Van Oldenborgh GJ, Otto FEL, Haustein K, AchutaRao K. The massive precipitation event of December 2015 in Chennai, India. BAMS. 2016;**97**:S87-S91

[12] Roxy MK, Ghosh S, Pathak A, Athulya R, Mujumdar M, Murtugudde R, et al. A threefold rise in widespread extreme rain events over Central India. Nature Communications. 2017;**8**(1):1-11. DOI: 10.1038/s41467-017-00744-9

[13] Dottori F, Szewczyk W, Ciscar JC, Zhao F, Alfieri L, Hirabayashi Y, et al. Increased human and economic losses from river flooding with anthropogenic warming. Nature Climate Change.

2018;**8**(9):781-786. DOI: 10.1038/s41558-018-0257-z

[14] Skilodimou HD, Bathrellos GD, Chousianitis K, Youssef AM, Pradhan B. Multi-hazard assessment modeling via multi-criteria analysis and GIS: A case study. Environmental Earth Sciences. 2019;**78**(2):1-21. DOI: 10.1007/s12665-018-8003-4

[15] The National Research Council. New Strategies for America's Watersheds. Washington, DC: National Academy Press; 1999

[16] Chowdary VM, Ramakrishnan D, Srivastava YK, Chandran V, Jeyaram A. Integrated water resource development plan for sustainable management of Mayurakshi watershed, India using remote sensing and GIS. Water Resources Management. 2009;**23**(8):1581-1602. DOI: 10.1007/s11269-008-9342-9

[17] The Saudi Geological Survey (SGS) official portal. [online] Available from: https://www.sgs.org.sa [Accessed: February 25, 2018]

[18] Maerz NH, Youssef AM, Pradhan B, Bulkhi A. Remediation and mitigation strategies for rock fall hazards along the highways of Fayfa Mountain, Jazan region, Kingdom of Saudi Arabia. Arabian Journal of Geosciences. 2015;**8**(5):2633-2651. DOI: 10.1007/s12517-014-1423-x

[19] General Authority for Statistics. 2018. Available from: https://www.stats.gov.sa

[20] Abdelkarim A, Gaber AF, Youssef AM, Pradhan B. Flood hazard assessment of the urban area of Tabuk City, Kingdom of Saudi Arabia by integrating spatial-based hydrologic and hydrodynamic modeling. Sensors. 2019;**19**(5):1024. DOI: 10.3390/s19051024

[21] Youssef AM, Pradhan B, Sefry SA. Flash flood susceptibility assessment in Jeddah city (Kingdom of Saudi Arabia) using bivariate and multivariate statistical models. Environmental Earth Sciences. 2016;**75**(1):1-16

[22] Foster S. Groundwater: Assessing Vulnerability and promoting protection of a threatened resource. Proceedings of the 8th Stockholm Water Symposium, Sweden. 10-13 Aug, 1998. pp. 79-90

[23] Scanlon BR, Keese KE, Flint AL, Flint LE, Gaye CB, Edmunds WM, et al. Global synthesis of groundwater recharge in semiarid and arid regions. Hydrological Processes: An International Journal. 2006;**20**(15):3335-3370. DOI: 10.1002/hyp.6335

[24] Chen J, Qian H, Gao Y, Wang H, Zhang M. Insights into hydrological and hydro chemical processes in response to water replenishment for lakes in arid regions. Journal of Hydrology. 2020;**581**:124386. DOI: 10.1016/j.jhydrol.2019.124386

[25] Chen J, Gao Y, Qian H, Ren W, Qu W. Hydro geochemical evidence for fluoride behavior in groundwater and the associated risk to human health for a large irrigation plain in the Yellow River Basin. Science of the Total Environment. 2021;**800**:149428. DOI: 10.1016/j

[26] Benz UC, Hofmann P, Willhauck G, Lingenfelder I, Heynen M. Multi-resolution, object-oriented fuzzy analysis of remote sensing data for GIS-ready information. ISPRS Journal of Photogrammetry and Remote Sensing. 2004;**58**(3-4):239-258. DOI: 10.1016/j.isprsjprs.2003.10.002

[27] Dutta D, Herath S. Trend of floods in Asia and flood risk management with integrated river basin approach. In: Proceedings of the 2nd International

Conference of Asia-Pacific Hydrology and Water Resources Association. Singapore. Vol. 1. 2004. pp. 55-63

[28] Gaitan S, Ten Veldhuis MC, van de Giesen N. Spatial distribution of flood incidents along urban overland flow-paths. Water Resources Management. 2015;**29**(9):3387-3399. DOI: 10.1007/s11269-015-1006-y

[29] Chen J, Wu H, Qian H, Gao Y. Assessing nitrate and fluoride contaminants in drinking water and their health risk of rural residents living in a semiarid region of Northwest China. Exposure and Health. 2017;**9**(3):183-195. DOI: 10.1007/s12403-016-0231-9

[30] Naiman RJ, Bisson PA, Lee RG, Turner MG. Approaches to Management at the Watershed Scale. In: Kohm KA, Franklin JA, editors. Creating a Forestry for the 21st Century: The Science of Ecosystem Management. Washington, D.C., USA: Island Press; 1997. pp 239-253

[31] Wilson JP, Gallant JC. Digital Terrain Analysis. In book: Terrain Analysis: Principles and Applications. Canada: John Wiley & Sons Inc.; Jan 2000. pp 1-27

[32] Mitas L, Mitasova H, Brown WM, Astley M. Interacting Fields Approach for Evolving Spatial Phenomena: Application to erosion Simulation for Optimized Land Use. USA: lllinois University at Urbana Center for Supercomputing Research and Development; 1996

[33] Mahmoud SH, Alazba AA. Integrated remote sensing and GIS-based approach for deciphering groundwater potential zones in the central region of Saudi Arabia. Environmental Earth Sciences. 2016;**75**(4):1-28. DOI: 10.1007/s12665-015-5156-2

[34] El-Hames AS. Determination of groundwater availability in shallow arid region aquifers utilizing GIS technology: A case study in Hada Al-Sham, Western Saudi Arabia. Hydrogeology Journal. 2005;**13**(4):640-648. DOI: 10.1007/s10040-004-0323-7

[35] Madani A, Niyazi B, Elfakharani A, Osman H. The effects of structural elements on groundwater of Wadi Yalamlam, Saudi Arabia using integration of remote sensing and airborne magnetic survey. Earth Systems and Environment. 2019;**3**(2):301-312. DOI: 10.1007/s41748-019-00091-3

[36] Swain S, Mishra SK, Pandey A, Pandey AC, Jain A, Chauhan SK, et al. Hydrological modelling through SWAT over a Himalayan catchment using high-resolution geospatial inputs. Environmental Challenges. 2022;**8**:100579. DOI: 10.1016/j.envc.2022.100579

Chapter 7

Ecological Status of Water Quality

Erjola Keci

Abstract

This chapter gives an overview of environmental quality standards and water quality elements according to the Water Framework Directive and all related guidance documents, Environmental Quality Standards Directive and Water Information System for Europe – WISE GIS Guidance. The chapter explains the main definitions and principles by which surface water and groundwater quality is assessed; it sets out the concepts and procedures that closely drive the development of the aquatic environment protection strategy for the basins for the purpose of preventing the damage to water bodies, as well as protecting, increasing, and rehabilitating the status of all bodies of water, both surface and groundwater ones.

Keywords: ecological status, biological quality elements, biological quality ratio, environmental quality standards, reference conditions, chemical status, quantitative status, water framework directive, heavily modified waterbody, artificial waterbody

1. Introduction

One of the requirements of Water Framework Directive is that each River/Lake Basin area should have a Management Plan, which needs to be updated each 5–6 years. All River Basin Management Plans implementing the Water Framework Directive in the European Union have as their overriding purpose the maintenance and protection of the aquatic environment. This is achieved through measures to ensure that all waters (surface and groundwater) are of sufficient and sustainable quality and quantity for both environmental and economic needs.

Article 4 of the Water Framework Directive (WFD) [1] defines the core concept of the Directive and the specific purpose of RBMPs, namely, to implement measures as appropriate to:

- Prevent deterioration of the status of all surface water and to achieve good ecological status or good ecological potential.

- Progressively reduce pollution from priority substances and river basin specific pollutants to achieve good chemical status for surface waters.

- Prevent or limit the discharge of pollutants generally to groundwater and reverse negative trends.

IntechOpen

- Prevent deterioration of the status of all groundwater, determined by quantitative status and chemical status, and to achieve good chemical status for all groundwater bodies.

- Ensure a sustainable balance of groundwater abstraction against annual recharge.

The European Union (EU) Water Framework Directive (WFD) [1] requires EU member states to set up monitoring and ecological quality classification systems in order to assess the ecological status of surface waters and gauge the extent of human impact on these water bodies. Inland, transitional, and coastal surface waters as well as groundwaters are covered by the WFD. By regulating specific pollutants and establishing corresponding regulatory standards, the directive ensures an integrated approach to water management while respecting the integrity of entire ecosystems.

The reference condition technique, which is what the Directive uses, classifies ecological state by comparing it to reference circumstances. Finding reference conditions, which are understood as the set of conditions to be anticipated in the absence of or under minimum anthropogenic perturbations, is a critical component of the ecological evaluation.

2. Ecological status

The quality of the structure and efficiency of aquatic ecosystems is described by an indicator called the Ecological State of Surface Waters. The ability of the water body to maintain well-organized and balanced animal and plant communities, which are essential biological instruments for supporting the self-purification processes of water, is interpreted as the ecological quality target set forth by Directive 2000/60/EC [1]. Using the physical-chemical and hydro-morphological aspects (such as the water regime and the morphological natural characteristics of the river body) as support for the process of establishing environmental quality, the legislation defines the ecological condition through the research of particular aquatic biological groups. In a high ecological state, or when anthropogenic activities are absent or minimal, the reference conditions define the needs of the biological quality elements. When anthropic activities are absent or of minor significance, as is the case in high ecological states, the reference circumstances define the needs of the biological quality elements. The primary distinction between reference conditions and high ecological status is that the former primarily describe biological quality components, whilst the latter additionally contain chemical-physical, hydro-morphological, and other components (see WFD, appendix II, 1.3) [1]. As a result, particular physical-chemical parameters are established for each kind of transition environment for each form of surface water body, which corresponds to the circumstances of the same in a highly ecological state.

According to Guideline n.13 "Classification of Ecological Status" [2], the degree of correspondence between the present and unaltered physic-chemical conditions only influences the transition between ecological status from high to excellent and from good to sufficient. As a result, the ecological state of the water body will be deemed sufficient if the biological quality components meet the requirements for excellent status, but one or more chemical-physical quality elements do not. As stated in Annex V, 1.1.3 of the WFD [3], both generic elements and particular pollutants are included in the chemical-physical elements.

The WFD in this regard strongly emphasizes the importance of using biological indicator elements (plankton, fish and benthic flora and fauna) in order to assess the ecological quality status and thus maintain the ecological health conditions. These indicators respond in an anticipated manner to anthropogenic disturbances and stressors and make possible long-term classification of ecological quality based on a predictable interaction between pressures and indicators. During the past decades, intensive efforts have been carried out in Europe to standardize and develop biological indicator metrics that can be used rapidly and on a large scale across the continent. WFD gives a framework for developing such scientifically sound and practicably applicable assessments for freshwater ecosystems. Linkage of biological indicators to pressures has been made possible by using a large set of multimeric indices. For fish communities, the multimeric index IBI (Index of Biological Integrity), for benthic macroinvertebrates indices such as Biological Monitoring Working Party (BMWP) and *Ephemeroptera*, *Plecoptera*, *Trichoptera* (EPT) as well as diversity indices, for diatoms several metrics are used, including Trophic Diatom Index (TDI). Many other locally developed indices are also used in specific countries to accommodate the knowledge and the existing set of the existing data.

The Directive stipulates that the ecological quality classification "... shall be represented by lower of the values for biological and physic-chemical monitoring results for the relevant quality elements..." (WFD, Annex V, 1.4.2) [4]. Based on this set of biological indicators the ecological status processing procedure involves, the calculation of the metrics expected for the quality elements monitored and the integration of the results at the water body level. The class of the ecological status of the water body will derive from the lowest class attributed to the different quality elements. The quality, expressed in five classes, can vary from high to bad. The worst ratings (poor and bad) are determined only by EQBs (biological quality elements).

The reference conditions are used to define the limits of the indicators, as shown in **Figure 1**.

The biological quality elements at good ecological status or potential must be supported, relevantly, by the established ranges and levels for the physic-chemical quality element.

It is fair to infer that the ranges and levels defined for the physic-chemical quality components should also be type-specific as the values for the biological quality

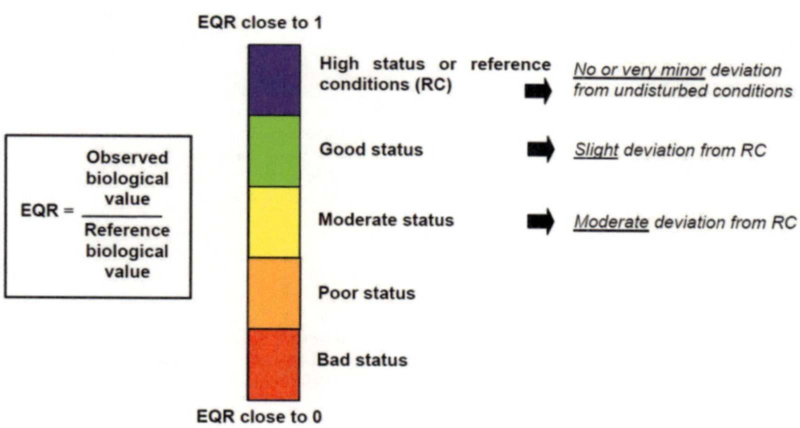

Figure 1.
Link between reference condition and limit definition [5].

elements at excellent status will be type-specific. The ranges or values for any or all of the physicochemical quality factors may be the same for several varieties.

Cases (a) and (b), in **Figure 2**, should be recognized, but for practical reasons, it is proposed to establish only one range or level including both aspects [6].

2.1 Chemical status

The legislation of the European Union includes safeguards against chemical contamination of surface waterways. Two parts are present. Choosing and regulating compounds of concern to the entire EU (the "priority substances" and "priority hazardous substances"); choosing by Member States substances of concern to their own countries or regions (the "river basin specific pollutants") for control at the appropriate level. The Environmental Quality Standards Directive (EQSD) 2008/105/EC (as amended by the Priority Compounds Directive 2013/39/EU) [7] establishes the relevant environmental quality standards (EQS) for surface waters, and surface waters are said to be in good chemical condition if their concentrations of "priority substances" [1, 7] do not exceed those EQS. EQSs seek to safeguard both human health and the most vulnerable aquatic animals from direct toxicity. Hazardous compounds must be eliminated for groundwater to achieve the goal of good chemical status.

2.2 Quantitative status

Whereas water quantity was regarded as an ancillary element to overall ecological status in the WFD at publication, water quantity in terms of flow regime is now seen as one of the primary elements of good ecological status [8, 9]. For rivers especially, specific assessments should be made to identify the 'environmental flow regime' and to quantify the degree of impact on the waterbody resulting from changes to the flow regime.

For groundwater, quantitative status is determined by a comparison of the rate of annual recharge of the groundwater stock against consumption of the groundwater. Abstractions exceeding recharge even in the short-term will negatively impact on

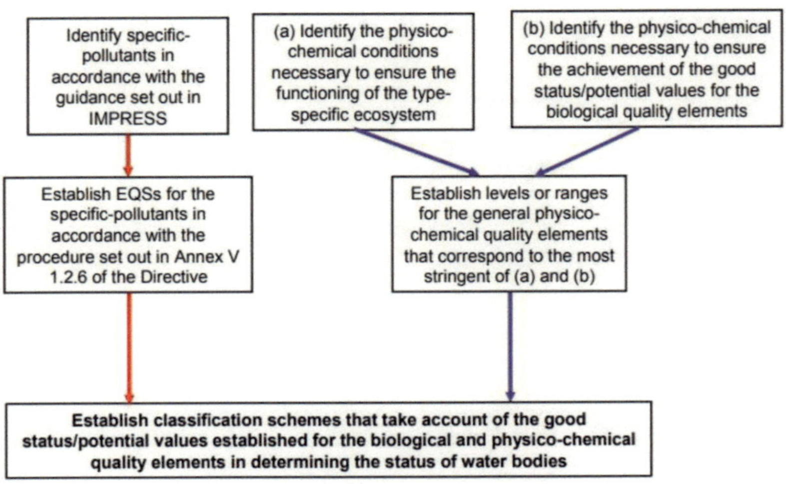

Figure 2.
The establishment of ranges and levels for the physic-chemical quality elements at a good ecological status potential [6].

local groundwater level with consequences for (a) surface water flows and ecosystems (b) availability of groundwater for economic uses generally.

2.3 Levels of status

The concept of waterbody status is central to the environmental deliverables of the WFD generally and RBMPs specifically. Status is seen as a relative concept, relative to the 'reference condition' which is always type-specific.

The fundamental concepts to keep in mind in the process of identifying the reference conditions can be summarized as follows:

- RCs do not necessarily equal undisturbed conditions; they can include small disturbances, i.e., they can include those anthropic pressures that do not cause ecological effects, or that induce negligible ecological effects;

- RCs equal high ecological status;

- RCs are represented by values of the ecological quality elements relevant in the status classification.

Typically:

- High Status - - No or very minor anthropogenic alterations to the biological quality, physic-chemical and hydro morphological elements of the waterbody.

- Good Status - - Slight levels of distortion of the composition and abundance of biological quality elements, with physic-chemical and hydro morphological conditions consistent with the achievement of good biological quality.

- Moderate Status - - Modest deviation of biological quality elements relative to High Status, with physic-chemical and hydro morphological conditions consistent with the achievement of moderate biological quality.

- Poor Status - - Waters showing evidence of major alterations to the values of the biological quality elements.

- Bad Status - - Waters showing evidence of severe alterations to the values of the biological quality elements.

For groundwater, status is defined only by the lesser of quantitative status and chemical status, thereby being either:

- 'Good' -

- or 'Fail' -

Figure 3 illustrates various components associated with an ecological status assessment generally. The comprehensive list of all quality elements is shown in Technical Appendix 1. Biological assessment results need to be expressed using a numerical scale between 0.00 and 1.00, the 'Ecological Quality Ratio' (EQR) [11].

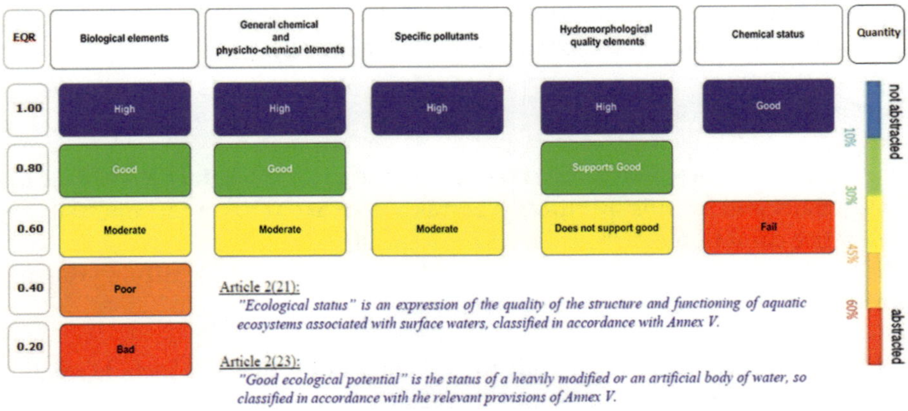

Figure 3.
Quality elements in waterbody ecological status [10].

An EQR value = 1.00 represents type-specific reference conditions (i.e., fully natural conditions). Values close to 0.00 = Bad Ecological Status (BES) [11].

The objective of the EQR is to ensure comparability between different assessment methods i.e., to provide a common scale of ecological quality between different river basins. The EQR intervals shown in **Figure 3** are indicative only and may vary depending on the BQE under assessment.

Flow abstraction has a direct impact on ecological status, even in modest quantity. **Figure 3** shows that typically, permanent, or very extended abstraction quantities >30% of the mean flow are likely to result in the hydro morphological quality element achieving only Moderate status [8]. Due to the importance of flow regime generally, it is probable that the ecological status of the waterbody would also be classed as Moderate.

2.4 Environmental objectives for surface waterbodies

To progress the determination, the status of individual waterbodies must be compared to a 'type-specific reference condition'.

For each surface waterbody type, type-specific biological, hydro morphological and physic-chemical conditions should be established representing values of the elements specified for that surface water body type at high ecological status [12]. 'Status' is therefore a condition relative to the reference condition, generally defined as the waterbody type in 'high status'.

Determination of the biological reference condition for each waterbody type is a matter for national/country procedures, although some EU standard criteria are fixed, especially with respect to 'priority substances' as set out in Directive 2008/105/EC [13].

2.4.1 Reference conditions for biological quality elements

Numerous approaches have been used in Europe to assess surface water biological quality conditions by evaluating taxonomic data. The general approach is to use the composition (defined as diversity + abundance) of the community at each site to infer water quality conditions. Some macro-invertebrates tend to be tolerant of poor water quality conditions e.g., order *Diptera* and the class *Oligochaeta*. Other organisms, for example, the orders *Ephemeroptera*, *Plecoptera*, and *Trichoptera*, are more sensitive to pollution.

Biotic index score range	Equivalent EQR	Expected water quality status	Degree of organic or chemical pollution	Degree of hydromorphological disruption[1]
e.g., 0.00–4.50	e.g., > 0.8	e.g., HIGH	e.g., slight	e.g., none
4.51–5.50	0.8–0.6	Good		e.g., slight
	0.6–0.4	Moderate		Moderate
	0.4–0.2	Poor		High
	< 0.2	Bad		Severe

[1]*Hydro morphological disruption encompasses physical habitat destruction, longitudinal discontinuity and/or significant disturbance to the hydrological flow regime through e.g., abstraction, mining, or hydropeaking, and can be objectively quantified through the use of European Standard EN 15843 [14].*

Table 1.
Relationship between biotic index and BQE ecological quality ratio.

Specific to the type of waterbody sensitive taxa are expected to be found only at sites with 'good water quality', varying according to specific national conditions. It can be generally inferred that if any of the specific 'quality elements' of waterbody status i.e., physic-chemical, specific pollutants, hydro morphological and/or priority substances are significantly adverse, then the biological quality elements (BQEs) can be expected to show a lower ecological quality ratio, as per **Figure 1** and **Table 1**.

The hydro-morphological, physic-chemical, and biological quality components represent undisturbed or nearly undisturbed circumstances, therefore strictly speaking the normative definition supplied by the WFD needs the development of reference conditions in relation to the values of the quality elements. At first glance, it would seem that the limitations of these quality element values may serve as the foundation for the reference requirements for the biological quality element. These values are unknown, nevertheless, until the development of reference circumstances. Hence, the crucial element in defining reference conditions for biological quality elements is the understanding of the range of values of the biological quality elements in healthy (entirely undisturbed) ecosystems and the amount of deviation allowed.

To forecast modest (acceptable) levels of change in the hydro-morphological, chemical, and physicochemical factors supporting the biological elements, criteria must be developed. The criteria that will be used to anticipate unimpaired circumstances need not be the same for various taxonomic groups due to the significant ecological differences between the taxonomic groups that may be utilized in monitoring operations (e.g., fish, phytoplankton, etc.). However, it must be ensured that the characteristics listed below, each of which can be connected to one or more pressure types, are of utmost significance when characterizing and maintaining biological reference conditions and then comparing various locations to these sites. These elements are all included in Annex V 1.2.1 of the WFD [1, 15].

Hydro morphological elements supporting the biological elements:

- hydrological regime
- quantity and dynamics of water flow
- connection to ground water bodies

- river continuity
- morphological conditions
- river depth and width variation
- structure and substrate of the riverbed
- structure of the riparian zone

Chemical and physicochemical elements supporting the biological elements:

- General
  - thermal conditions
  - oxygenation conditions
  - salinity
  - acidification status
  - nutrient conditions
- Specific pollutants
  - pollution by all priority substances identified as being discharged into the body of water.
  - pollution by other substances identified as being discharged in significant quantities into the body of water.

The WFD defines the following biological characteristics for use as indicators of ecological state for the quality element "fish": species composition, abundance, sensitive species, age structure and reproduction.

Several metrics are used all over Europe to describe these elements. For benthic invertebrate fauna beside taxonomic composition, the abundance and ratio of disturbance sensitive taxa to semi sensitive and tolerant ones are taken into consideration. For macrophytes and Phyto-benthos taxonomic composition and average abundance are taken into consideration. Reference conditions, representing biological communities in high status waterbodies, should be defined for each individual metric, and for each area and river type. An example of Biological Quality Element is given herein below (**Figure 4**).

The European Water Framework Directive (WFD) [1] mandates that an intercalibration process be used to reconcile national classifications of excellent ecological state. Because different geographical areas in Europe harbor different conditions in terms of biological indicators, geography, substrate and climate.

In this effort, the good status limits of the national evaluation techniques are compared and, if required, adjusted in order to harmonize the major disparities in status categorization among the states. Distinct freshwater categories are intercalibrated,

(a) Example Biological Quality Element	(b) Example (type-specific) conditions specified for the element at good status	(c) Examples of indicative parameters (metrics) based on measurements of composition and abundance	
Benthic Invertebrate Fauna (rivers)	THERE MUST BE NO MORE THAN SLIGHT CHANGES IN COMPOSITION AND ABUNDANCE THERE MUST BE NO MORE THAN SLIGHT CHANGES IN THE RATIO OF DISTURBANCE SENSITIVE TAXA TO INSENSITIVE TAXA THERE MUST BE NO MORE THAN SLIGHT SIGNS OF ALTERATION TO THE LEVEL OF DIVERSITY	Presence or absence of particular species or groups of species Overall richness or richness of particular taxonomic groups Relative number of taxa in particular taxonomic groups Abundance of particular species or groups of species Relative abundance of particular species or groups of species Overall diversity, or diversity within particular taxonomic groups	Taxa could be selected and/or grouped by known sensitivity/tolerance, feeding type, habitat preferences, etc

Figure 4.
Example of benthic invertebratefauna as one of biological quality elements.

with an emphasis on a few different types of water bodies (intercalibration types), human stresses, and biological quality elements. The WFD Common Implementation Strategy Guidance paper on the intercalibration process outlines the steps that must be followed when conducting intercalibration exercises. These exercises are conducted in Geographical Intercalibration Groups, which are bigger geographical entities made up of Member States with comparable water body types.

According to the WFD/Annex II, 1.3 [12], the reference sites and conditions are designated based on the following options and requirements:

- Spatially based reference conditions using data from monitoring sites;
- Reference conditions based on predictive modeling
- Temporally based reference conditions using either historical data or paleo-reconstruction or a combination of both;
- A combination of the above approaches.

In cases when it is not possible to imply the abovementioned methodology, according to the WFD, reference conditions can be established based on the expert judgment taking into consideration the country specificities.

2.4.2 Reference conditions for physic-chemical elements

Together with knowledge about water users and site-specific elements, water-quality criteria serve as a starting point for setting water-quality targets. Information is acquired, among other things, based on the following to determine water-quality objectives:

- Inventories of current and potential new water uses;
- Inventories of emission sources including point and non-point sources, and of sites of production, use, storage and disposal of hazardous substances which could accidentally be emitted into aquatic ecosystems;

- Results of water-quality monitoring and/or water-quality assessments and classifications;
- Surveys of specially protected waters such as drinking-water reservoirs and groundwater, and specially protected areas such as wetlands;
- Results of hydrological measurements and related information (e.g., run-off, hydraulic characteristics of water bodies).

Assessments and classifications of water quality are often based on a reference system that is used to measure and compare the present water quality. A reference base's objective is to characterize, as accurately as feasible, the natural characteristics of a water body or the fundamental demands that various water users place on water quality. The establishment of such a reference base presents various challenges, particularly concerning watershed regions that have experienced anthropogenic stress for decades and the use category "aquatic life." These days, a lot of nations utilize water quality standards for that. To develop a suitable reference basis, other nations conduct detailed hydro-chemical examinations of the relevant catchment region. However, establishing a reference standard for all water-quality metrics is almost impossible.

2.4.3 Reference conditions for priority hazardous substances

Good chemical status is attained for a water body when it conforms with the EQS for all priority compounds and other pollutants specified in Annex I of the EQSD, in accordance with WFD Annex V, 1.4.3 [16] and the EQSD Article 1 [7]. The total concentrations in the entire water sample are how the water environmental quality standards (EQS) are stated. When it comes to cadmium, lead, mercury, and nickel, however, the water EQS relates to the dissolved concentration, or the dissolved phase, of a water sample that was produced by filtering via a 0.45 m filter or other comparable pre-treatment.

When assessing the monitoring results against the relevant EQS, the following can be considered: natural background concentrations for metals and their compounds where such concentrations prevent compliance with the relevant EQS and hardness, pH, dissolved organic carbon or other water quality parameters that affect the bioavailability of metals.

2.4.4 Reference conditions for other specific pollutants

For assessment of the status of a surface water body regarding the other specific pollutants, 12 water samples per year (monthly sampling) should be taken at a representative monitoring point. For classification, the following principles apply:

1. The assessment is conducted per water body, using the monitoring data collected for one calendar year.
2. The 95-percentile value of the monitoring data is used for comparison with the environmental quality standards per pollutant and the results classify the water quality as:
    - high,

- good,
- moderate
- poor, and
- bad

3. The final class is determined by the poorest score for one or more individual parameters.

2.4.5 Reference conditions for heavily modified and artificial waterbodies

According to WFD Article 2(9) [1], there are two components to the definition of a HMWB. To be a HMWB a water body must be:

- physically altered by human activity
- substantially changed in character

A waterbody may only be designated as heavily modified if it has passed through the designation procedure involving both tests as specified under WFD Article 4(3) (a) & (b) [1]. The tests must be waterbody specific since they are intended to guarantee that HMWBs are only recognized in cases where there are no plausible chances for obtaining excellent status within a water body. The designation and the reasons for it must be specifically mentioned in the RBMP.

As for natural waterbodies, the environmental objectives for HMWBs and AWBs are defined relative to a reference condition, which is the maximum ecological potential (MEP) of the waterbody. The MEP is the condition where, taking into account the altered properties of the waterbody, the biological status most closely resembles that of the nearest similar surface waterbody. With regards to its biological status, Good Ecological Potential (GEP) accommodates "slight changes" from the MEP.

Once designated as HMWB or AWB, the environmental objectives are "good ecological potential" (GEP) and good chemical status. Because it accounts for the ecological effects of physical changes required to support a given purpose, such as flood protection or hydropower, GEP is a less strict target than GES.

2.4.5.1 Hydro morphological impacts of hydropower and HMWB designation

The objective determination of the extent of potential hydro morphological impact from HPPs is highly relevant with respect to waterbody status. The hydro morphological condition of the waterbody is integral to its overall status. As confirmed by EU CIS Guidance 31 [8], flow regime has a preeminent controlling effect on ecosystem health, and therefore waterbody status.

The relevance of hydropower operation is that very frequently, the downstream flow regime is entirely disrupted, either in terms of the magnitude of the flow abstracted relative to the mean flow at the point of abstraction, and/or the duration of time that the flow abstraction takes place. This flow regime disruption is additional to the mainly morphological (physical) impacts implied by EU CIS Guidance 4 [10].

An objective test to determine the level of impact of hydropower is essential within the context of Heavily Modified Waterbody (HMWB) designations.

Clearly, construction of major weirs, barrages or dams have a substantial physical impact on the waterbody, altering the longitudinal continuity and ecosystem integrity. The length of the depleted reach for offline HPP systems is a critical factor in ecosystem impacts. A depleted reach is defined as the length of the river waterbody between the point of abstraction and the point of flow discharge. Long depleted reaches arise when the hydropower Operator is seeking to maximize the hydraulic head (H) operating at the turbine, thus maximizing the power output. If the HPP system is offline, then this maximization of Head invariably means excessively long depleted reaches.

The deficient hydro-morphological status will apply for the full length of the depleted reach, albeit on a 'reducing balance' basis. At the point of hydropower discharge, the river flow is restored and for 'run of river' systems, equilibrium is restored.

However, for the worst-case design of a major Dam + an offline HPP system (i.e., HPP not at the Dam), then at the point of discharge (where turbine operation is a function of storage, not of river flow), the disrupted regime may then continue for a further significant distance downstream, as the turbines discharge significantly more than the expected natural flow, the degree of impact depending on the downstream hydrology.

A single large storage based offline HPP system may therefore significantly impact on e.g., 20 km + of waterbody, depleting flow in the upper reaches, and over-compensating in the lower reaches, and thoroughly disrupting if not completely destroying the aquatic ecosystem.

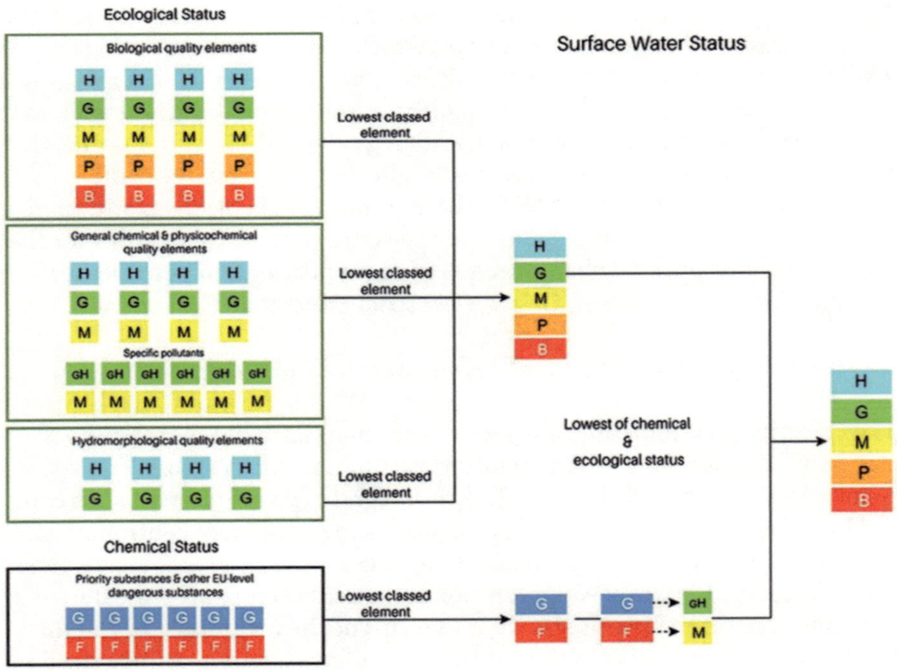

Figure 5.
Procedure for determination of waterbody overall status (European Commission-CIS Guidance 13 – Approach to the classification of ecological status) [6].

2.4.6 Final status determination of surface waterbodies

The final or overall status of each delineated waterbody is derived through a complex evaluation of elements shown in **Figure 5** [6]. According to WFD CIS guidance [6, 10], the final status of the waterbody should derive from the lowest classed element in each group.

2.5 Environmental objectives for groundwater bodies

The WFD requires Member States to designate separate groundwater bodies and ensure that each one achieves 'good chemical and quantitative status' (WFD Article 2(24) (25)) [1]. Determination of groundwater status is not as complex as those for surface water bodies. However, groundwater protection is the subject of numerous interacting Directives, including principally Directives 2006/118/EEC and 91/676/EEC

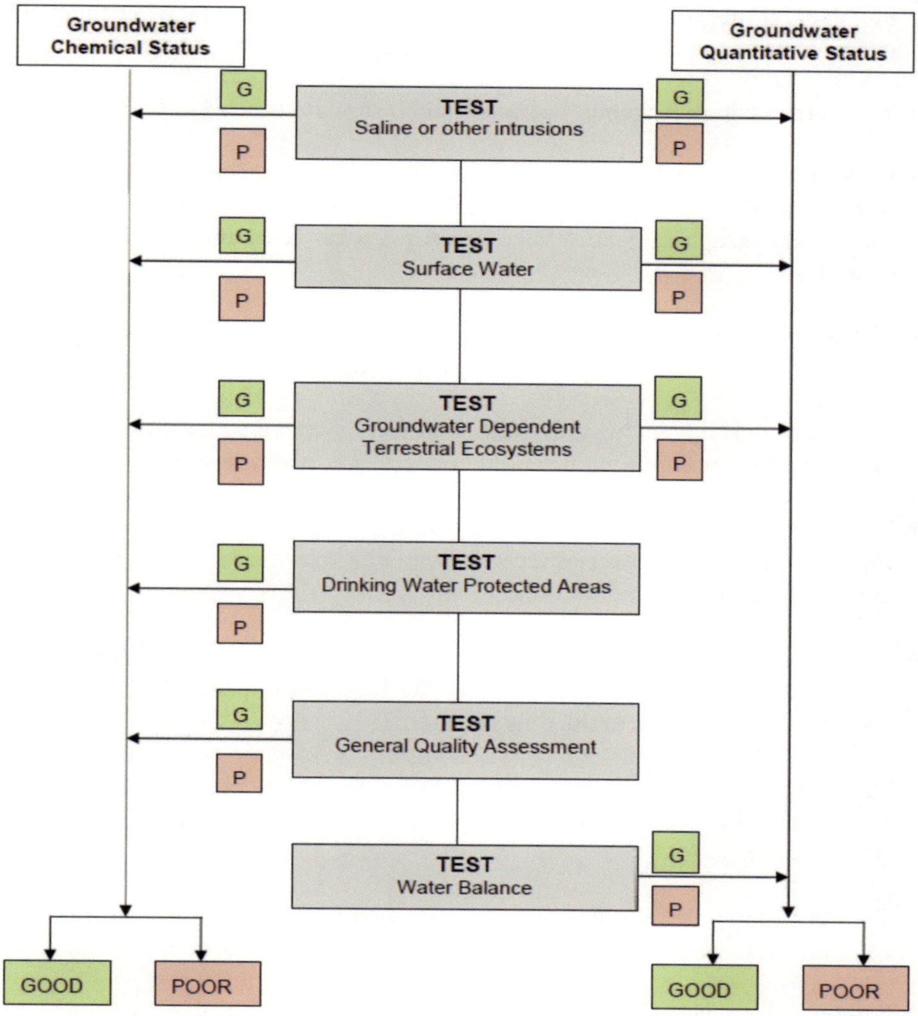

Figure 6.
Standard procedure for determination of groundwater overall status.

[17, 18]. Primarily through Directive 2006/118/EC [19], a management regime should be established which sets groundwater quality standards and introduces measures to prevent or limit inputs of pollutants into groundwater. Member States should establish standards at the most appropriate level and consider local or regional conditions [5].

The standard procedure for determination of the overall status of groundwater bodies is shown in the **Figure 6**.

3. Conclusions and recommendations

- According to WFD, EU member countries are obliged to prepare management plans for each River/Lake Basin area. The non-EU countries and those aiming to adhere to EU, depending on their status of pre-accession are recommended to align their national water related legislation with the EU legal framework.

- The EU pre-accession countries should harmonize and adopt EU water monitoring plans, strategies and practices for the periodical assessment of water quality, maintenance and achievement of good quality status of water.

- The water basin management plans should be updated each 5–6 years.

- Under the framework of integrated management of water resources, for each water basin should be identified a number of water bodies in different categories, based on geographical features and pressures/socio-economic activities within a water body.

- The countries should apply monitoring and ecological quality classification systems to assess the ecological status of waterbodies.

- Per each waterbody hydro-morphological, physic-chemical, chemical and biological quality component should be monitored and assessed.

- For the biological quality elements assessment, locally developed indices may be also used in specific countries accommodate the knowledge and the existing set of the existing data.

- For rivers especially, specific assessment should be made to identify the 'environmental flow regime' and to quantify the degree of impact on the waterbody resulting from changes to the flow regime.

- For groundwater, quantitative status is determined by a comparison of the rate of annual recharge of the groundwater stock against consumption of the groundwater.

- The definition of the ecological status per water body should be based on the comparison of real quality results with type specific reference condition.

- Reference condition should be defined per each group of quality elements. Reference conditions of a waterbody represent undisturbed or nearly undisturbed circumstances for that water body.

Acronyms and abbreviations

AWB	artificial waterbody
BES	bad ecological status
BQEs	biological quality elements
BMWP	biological monitoring working party
CIS	common implementation strategy
EPT	*Ephemeroptera*, *Plecoptera*, *Trichoptera*
EQR	environmental quality ratio
EQS	environmental quality standard
EQSD	environmental quality standards directive
EU	European union
GWB	groundwater body
GEP	good ecological potential
IBI	index of biological integrity
HMWB	heavily modified waterbody
HPP	hydropower plant
MEP	maximum ecological potential
PE	population equivalent
RC	reference conditions
RBMP	river basin management plan
SWB	surface water body
WFD	water framework directive
WISE	water information system for Europe

Author details

Erjola Keci
Faculty of Professional Studies, "Aleksander Moisiu" University, Durres, Albania

*Address all correspondence to: erjolakeci@yahoo.it

References

[1] European Commission–Water Framework Directive

[2] Common Implementation Strategy for the Water Framework Directive (2000/60/EC)

[3] European Commission–Water Framework Directive, Annex V, 1.1.3

[4] European Commission–Water Framework Directive, Annex V, 1.4.2

[5] Guidance document No. 10, River and lakes – Typology, reference conditions and classification systems, page 21

[6] European Commission-WFD CIS Guidance Document No. 13

[7] Environmental Quality standards Directive-EQSD, Annex II

[8] European Commission–CIS Guidance No 31—Ecological Flows in the Implementation of the WFD

[9] European Commission-Environmental Flows as a Tool to Achieve WFD Objectives-Discussion Paper, 2012

[10] Water Information System for Europe–WISE GIS Guidance – on the reporting of spatial data to WISE, v 6.0.6, 2016

[11] European Commission–Water Framework Directive, Annex V, 1.4.1

[12] European Commission–Water Framework Directive, Annex II, 1.3

[13] European Commission–Directive 2008/105/EC–The Environmental Quality Standards Directive

[14] European Standard EN 15843

[15] European Commission–Water Framework Directive, Annex V, 1.2.1

[16] European Commission–Water Framework Directive, Annex V, 1.4.3

[17] European Commission–Directive 91/676/EEC—concerning the protection of waters against pollution caused by nitrates from agricultural sources

[18] European Commission–Directive 2006/118/EC—on the protection of groundwater against pollution and deterioration

[19] European Commission–Groundwater legislative framework. Available from: https://ec.europa.eu/environment/water/water-framework/groundwater/framework.htm

Chapter 8

Integrated Water Clarification

Juhani Pylkkanen

Abstract

An integrated water clarification method for industrial and communal wastewater treatment and recycling is presented here. The clarification is hermetic and consists of four seamless phases: (1) separation of dissolved ingredients, (2) molecular activation, (3) clarification reactions and (4) replacement dissolution. It separates dissolved gases such as radon, carbon dioxide, hydrogen sulfide and hydrocarbons and dissolved solids such as iron, manganese, calcium, fluorine and phosphorus. The molecules are activated, and clarification reactions occur immediately by the suction of clean air, oxygen or ozone. The clarified water is aerated or oxygenated immediately after clarification. All this happens within a second or a few seconds depending on the water volume to be clarified. The integrated clarification is based on the fact that water itself is pure; it just picks up and easily transfers heavy loads. In the natural water cycle, the water eliminates its load by vaporizing. The integrated clarification separates the load in such a way that most of it can be removed and recycled. Treated water can be recycled and discharged into nature with respect to its natural value. The treated water can influence the natural water cycle by being sprayed on the ground and plants for solar vaporization and cooling of the climate. The load of the ocean can be reduced significantly. Integrated water clarification is performed under flowing conditions in a tube by kinetic energy. Gases and possible additives are sucked by a vacuum caused by a symmetrical nozzle design of two curved flow channels. It reduces water, energy and chemical consumption, and the capital and operational costs are low.

Keywords: wastewater clarification, wastewater treatment, water purification, water recycling, particle separation, pharmaceutical removal, radon removal, water disinfection

1. Introduction

The present wastewater treatment needs a new perspective and environmental respect. In the present water cycle, seas and oceans are rubbish dumps. The loads of rain and wastewater are pumped as quickly as possible into the rubbish dumps through the large underground sewer network. Climate cooling by water evaporation is reduced to a negligible level. There is little water and ingredient recycling. Despite the increase in power consumption and technology intensity in wastewater treatment, the environmental conditions indicate unsatisfactory results. Furthermore, the sewer network causes serious floods. Many serious problems, disasters and serious microbial growth can be avoided by water clarification and wide-area water buffers with solar evaporation.

The integrated water clarification process, design criteria, related devices and some practical case studies are presented here. This clarification consists of four seamless treatment phases, and as the fifth step, a firm gas bubble is generated as follows:

1. Separation of dissolved ingredients

2. Activation of molecules

3. Clarification reactions

4. Replacement dissolving

5. Bubble generation in water flow

There are still the following guiding principles for development and related research:

- Water itself is a pure substance that easily picks up heavy loads and can be purified by releasing the load.

- Water is present everywhere, and it performs the largest transportation system in the world.

- The clarification is to be well controlled and hermetic.

- Water purification can be combined with other water processes, such as hydro power and fountains.

- Treated wastewater is discharged into the natural water cycle via climate and natural wellness

- Water recycling is maximized in use with the lowest energy consumption and heat generation.

- Treated wastewater can be reused for climate cooling, such as solar evaporation.

- The solution is to be economically competitive, and its life cycle costs are a concern.

- Humans have made many mistakes and malfunctions in the natural water cycle that should be identified and corrected.

- Natural energy, e.g., potential and direct solar energy, is utilized.

Figure 1 summarizes the development principles. Integrated clarification fulfills these development and design principles in several areas, but further development and research are needed for many challenging applications. The separation of dissolved ingredients is effective, but removing these components is challenging for certain substances, such as chlorine and salt.

The development is performed by small innovative companies, Hilla Consulting Oy, T_2O Consulting Oy and T_2O Consulting GmbH. Application development and sales have been completed together with SansOx Ltd. Several universities have

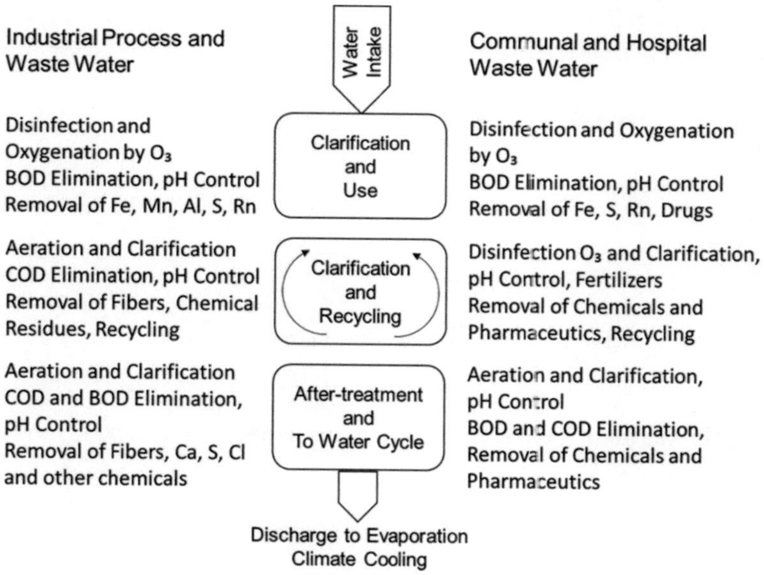

Figure 1.
Water recycling is integrated into the natural water cycle. There are several recycling and evaporation systems. Water picks up heavy loads that are to be released where it is caused, and there are only two acceptable options—Recycling or evaporation—For the water cycle.

contributed to related research activities. T2O Consulting GmbH has manufactured functional and visible prototypes and tested them according to several thousand integrated clarification criteria.

2. Integrated water clarification

A new hermetic water treatment process and device called OxTube Integrated Clarification and some of its applications are briefly presented here. The focus is on wastewater clarification and recycling. However, integrated clarification is suitable for the treatment of all kinds of water matrices. The OxTube is the basic water treatment machine and module that can be modified easily for various tasks, water matrices and installations, and it fulfills the guiding development principles set in Chapter 1, which is the most important.

To clarify water all the way, the following main criteria are identified:

- Dissolved ingredients should be separated, and molecules should be activated first in the clarification process.

- Other molecules present should also be activated or filtered out before integrated clarification, depending on the case.

- The wastewater content of various solids and chemicals distributed unevenly should be considered in the early phase of the integrated clarification process.

- Clarification gases should be fed and mixed evenly to ensure a high probability of collisions between molecules.

- The clarification should be well controlled, and all kinds of disturbances should be eliminated in the clarification process.
- It is desirable to refresh the water after clarification reactions by oxygenation or aeration.
- The removal of gases and particles should be considered after clarification.
- A dirt-free system should be considered.

The following solutions for several alternatives were developed after analysis and test runs:

- The nozzle was identified as the key element. Hundreds of nozzle models were produced and tested.
- A nozzle with curved water channels was identified as the best solution for separation, molecular activation, gas suction and mixing.
- The curved channels generate a pressure difference in the water flow divided into two channels, and a vacuum pocket is created in the middle of the nozzle zone.
- The pressure difference and vacuum pocket separate and activate ingredients dissolved in the intake water.
- The nozzle structures of two or more curved channels must be accurately symmetrical.
- **Figure 2** shows the visible efficiency of the separation and activation as well as the clarification with Swiss Alpen household water.

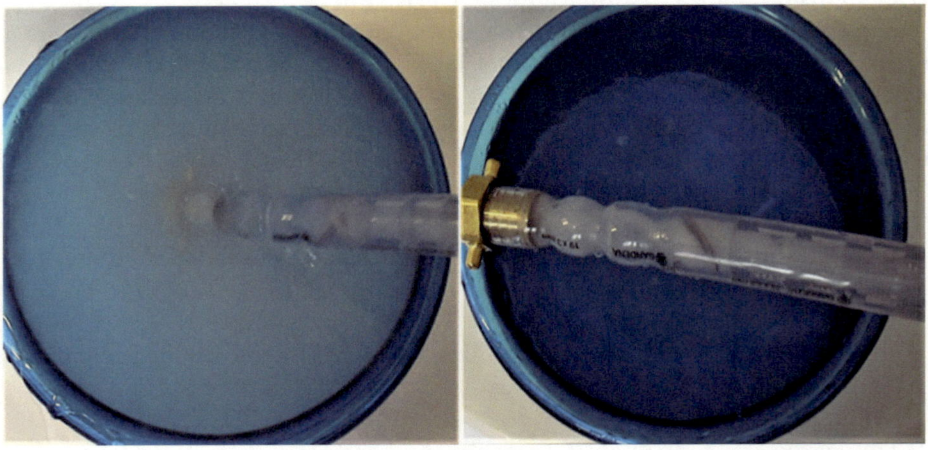

Figure 2.
Efficiency of OxTube integrated clarification with air; Swiss Alpen tap water was used only for soluble ingredient separation and molecular activation, and the integrated clarification was performed under the same conditions. Separation, activation, clarification and replacement are performed seamlessly within 1 second or less in an OxTube [1].

The channels are designed in such a way that there is no discontinuity that might cause disturbances in the clarification and dirt sticking in the nozzle.

The vacuum pocket sucks gases and possible chemicals and mixes them effectively and evenly in the water flow. **Figures 3** and **4** illustrate the effective suction and even mixing. The mixture of water and air is evenly foggy, as shown in **Figures 2** and **3**. The spray of the mixture of water and air is firm and coherent, as shown in **Figure 4**. The air suction is great; it can be more than 10 times greater in volume than the water flow, and the gravity decreases. No additional energy is needed.

3. OxTube machine

To optimize the total solution according to the challenging customer needs and concerns of various water matrices, a modular structure of the system is applied. Various customized solutions with natural respect are built from the same modules. The OxTube Machine is the key module in integrated clarification and has low power consumption. The treatment does not require any additional energy, but side processes such as ozone and chemical generation and feed might require power. Some solutions installed in pressurized piping might require compressed air. However, it is possible to avoid this by the system design.

The OxTube Machine seamlessly performs the following clarification phases:

- Separation of the dissolved ingredients (**Figure 2**)
- Activation of molecules (bubbling of reaction gases in **Figure 2**)
- Clarification gases and possible chemicals (**Figures 3** and **4**)

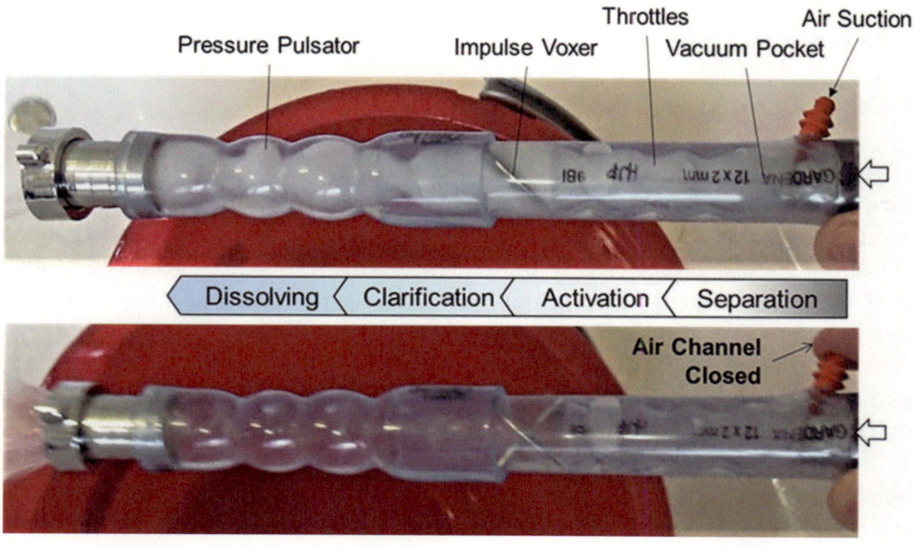

Figure 3.
The OxTube integrated water clarification was visualized with air suction. Integrated clarification is performed within less than a second [1].

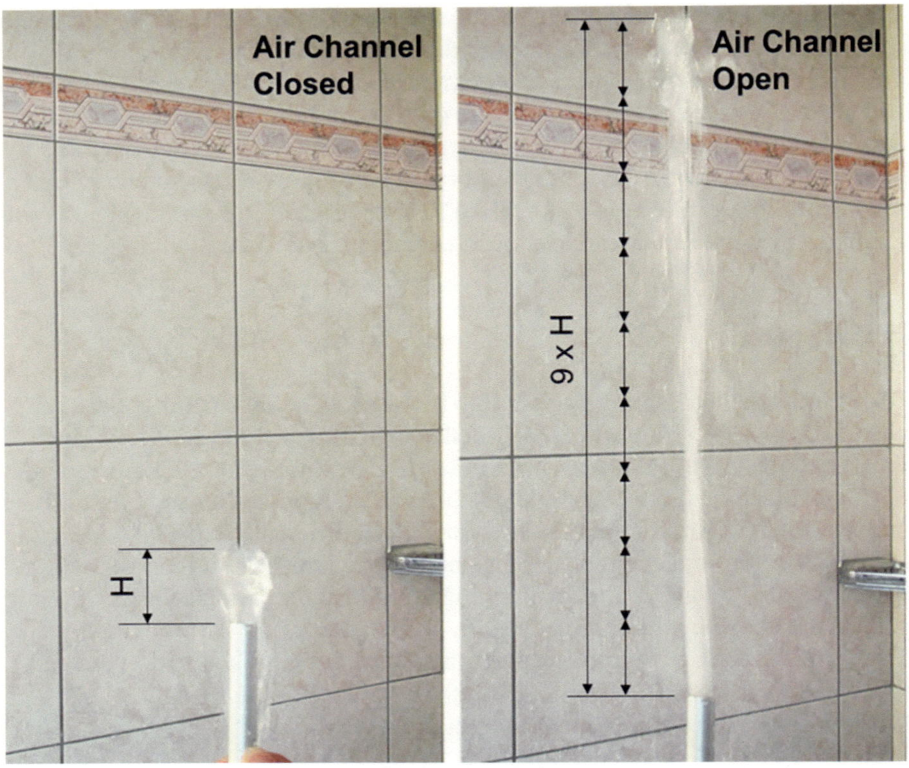

Figure 4.
The performance of the OxTube machine in gas suction and mixing, as well as in clarification and replacement dissolving, is great and efficient. No additional energy is needed in the integrated clarification. The height of the mixture is nine times greater than that of plain water under the exact same conditions of water intake.

- Performance of the clarification reactions (**Figure 2** right)
- Replacement dissolving
- Generation of gas bubbles (**Figure 4**)

The separation of the dissolved components and molecular activation are shown in **Figure 2** (left) when the water was removed through an OxTube without air suction. Separated gases cause smells and solid white particles. Molecular activation generates light bubbling and a gaseous smell. Immediate clarification can be achieved by air feeding into the activated water. Integrated clarification with air suction is completed within less than a second, as shown in the right panel of **Figure 2**.

The upper photo of **Figure 3** visually shows the even mixture of water, separated ingredients and air in the tube. The mixture is seen in white. Its gravity is much lower than that of the intake water matrix. The gravity reduction in the nozzle zone could be 1/10 that of water. The probability of collisions among molecules, particles and bubbles is extremely high. The lower photo of **Figure 3** shows the water flow when the air channel is closed. No turbulence or cavitation can be observed.

Bubble generation and mixing with water are visualized in **Figure 4**. The height of the outtake spray increases nine times when the air channel is open. The mixture gradually turns light in the nozzle. The spray of the mixture stays coherent all the way up, which is an essential feature of the OxTube Machine.

Figure 5 shows the oxygenation efficiency of the OxTube at an early stage development [2]. Since 2016, the separation of dissolved ingredients has improved, and gas suction has increased by a factor of 5. Losses are measured today in terms of mass flow due to large gravity drops and changes from incompressible to compressible liquids in the nozzle zone. The mass flow losses are measured to be less than 2%.

The OxTube Machine is presented in **Figure 6**. It consists of a DuOx nozzle, an air suction nipple, an outer tube and flanges. For small flanges, the flanges are replaced by standard thread fittings. All the fittings are metal-to-metal tight fits; there are no individual seals at all. Welding is minimized; only the flanges must be welded to the outer tube. The structure is very compact and easy to maintain in good condition.

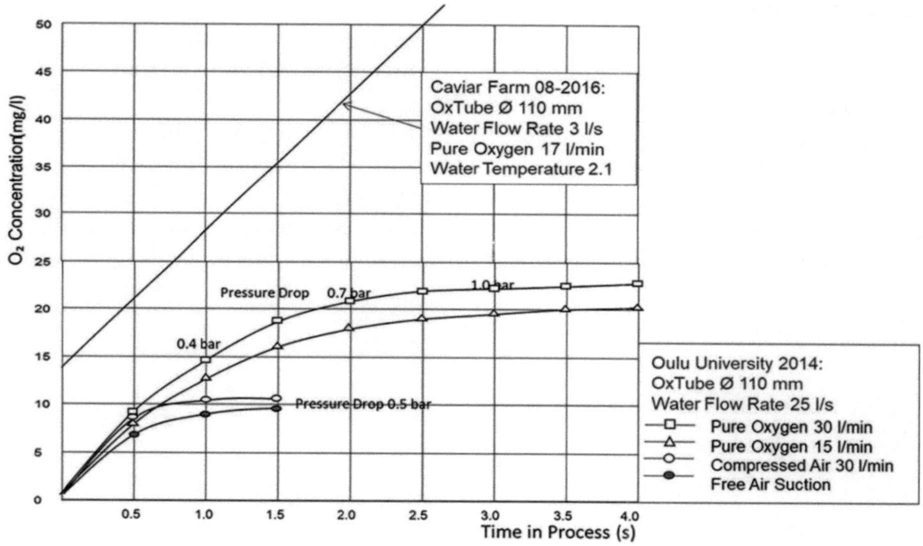

Figure 5.
Oxygenation efficiency of the OxTube in the early stage of development [2].

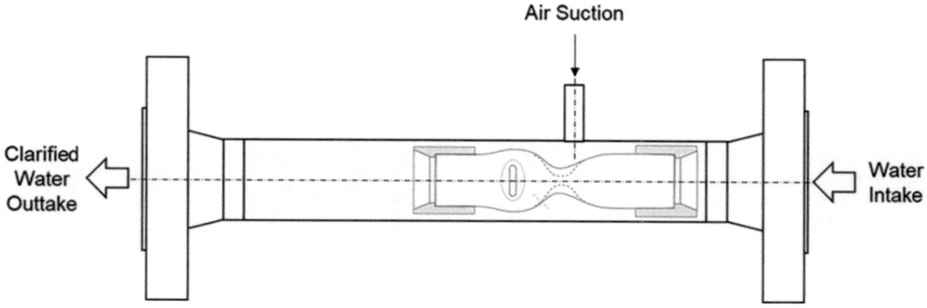

Figure 6.
The OxTube machine with gas and chemical suction is compact and includes standard fittings. It can be installed in various water treatment systems. The pressure norm is up to PN40.

OxTube machines are easy to modify and optimize for various water matrices and treatments. **Figure 7** shows the machine modified to realize heavy gas removal and water clarification. There is a booster nipple for pressured air feed that removes separated heavy gases such as radon and carbon oxide.

The two serial OxTube machines were combined so that the first step involved separation and activation, and the second step involved clarification, dissolution and bubble generation (**Figure 8**). Clarification performance is improved, particularly in very challenging cases. Furthermore, a booster can be applied in the second machine when heavy gases are involved in the integrated clarification.

There are many other ways to determine various combinations of the same modules to create the best solution to meet environmental and economic requirements. Clarification, disinfection and oxygenation of a water matrix can be combined in one OxTube by ozone or a mixture of ozone and air feed. Growing environmental problems, microbial variation and growth and drug residues are eliminated at the same time as water clarification.

Details of DuOx Nozzle are not presented here due to its dedicated structure that is to be kept in secrecy. It is the active element of the integrated clarification. Its functional features are presented above in this chapter.

An OxTube Machine separates dissolved ingredients, but removing them from the mixture requires further attention. An OxTube Machine generates a large amount of overdosed bubbles by gas suction, which can be used for flotation (**Figure 9**). The collision probability of particles and bubbles is high, and attachment is identified to be good, so flotation is a natural method of particle removal; it is similar to a built-in

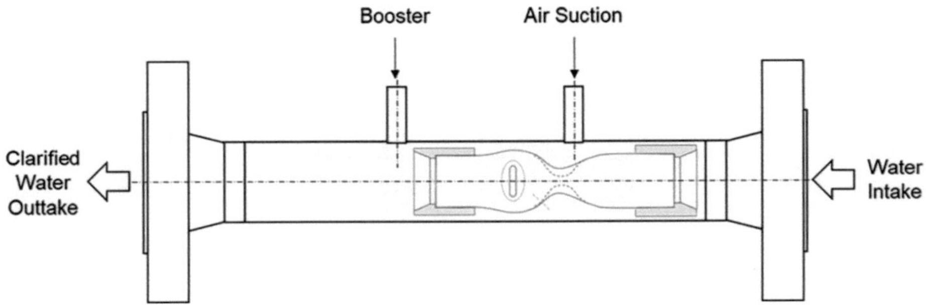

Figure 7.
OxTube machine with suction and booster nipple. The module is the same as that in **Figure 6** *and is composed of a booster nipple.*

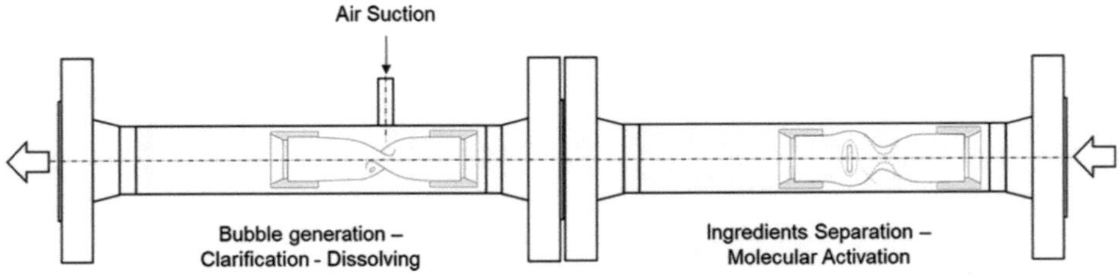

Figure 8.
With the combination of two OxTube machines, the first is responsible for separating the dissolved ingredients and molecular activation, and the second involves bubble generation, clarification and replacement dissolution. This is a solution for dirty waters such as wastewaters.

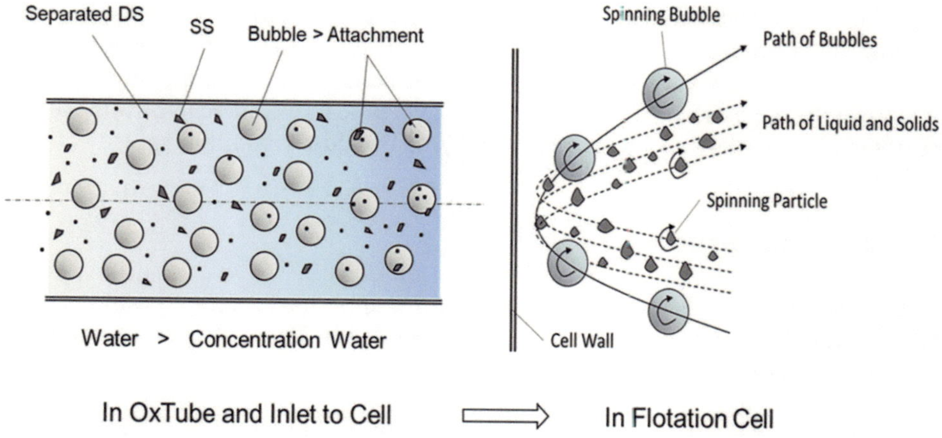

Figure 9.
The OxTube machine generates bubbles and even mixtures under hermetic tube conditions. The probability of particles and bubbles being met is high, and particle attachment already occurs in the OxTube and subsequent tubes before the flotation cell [1, 3, 4].

Figure 10.
OxTube tests of industrial wastewater resulted in a turbidity of 0.0 NTU. The clarified water is recyclable in industrial process use.

feature. Particle attachment occurs in the OxTube and Fitting Modules before the flotation cell, and the efficiency is improved. Compressed air and diffusers can be eliminated when an OxTube Machine is applied. Of course, all the filtering methods can be applied for particle removal after integrated clarification.

Figure 10 shows the efficiency of the integrated clarification according to one hundred real test runs completed with industrial wastewater. Turbidity of the clarified.

The amount of wastewater returned to 0.0 NTU within 30 minutes. Only an OxTube Machine, a pump and an open outtake vessel were used, and no filters were used. The treated water was identified as recyclable for industrial purposes. Recycling for household purposes needs further investigation and verification. The potential to address all of these water quality issues exists. Discontamination can easily occur with ozone, and the pH can be controlled with CO_2. Chemical residues can be removed by integrated clarification, as verified previously, or by additional filtering.

4. Water treatment unit

There are many methods and combinations for applying integrated water clarification technology and removing separated water loads. The basic water treatment unit called the WTU is presented in **Figure 11**. It consists of an OxTube Machine and a Gas Removal.

GasRemox and Fitting Module. It was developed for industrial process water, intake water from lakes and pretreated communal wastewater. All kinds of OxTube machines and modular combinations can be applied in the WTU. Only the most challenging water clarification cases are presented here.

Disinfections of various water matrices, including communal and hospital wastewater, can be disinfected by ozone instead of air. **Table 1** presents the results for the disinfection of hospital wastewater. A microbe reduction of 100% was achieved together with clarification and oxygenation of the over $20_{mg\ O2}/L$.

The efficiency of the OxTube system for the removal of pharmaceutical and chemical residues from city wastewater is presented in **Table 2**, which includes comparisons with two present systems. Pharmaceutical residues are a fast-growing disaster in natural water and ground. Approximately 90% of these residues were split, clarified and removed only by one run through an OxTube Machine to an open vessel without any filters. Two runs result in a 99 percent reduction. The city wastewater

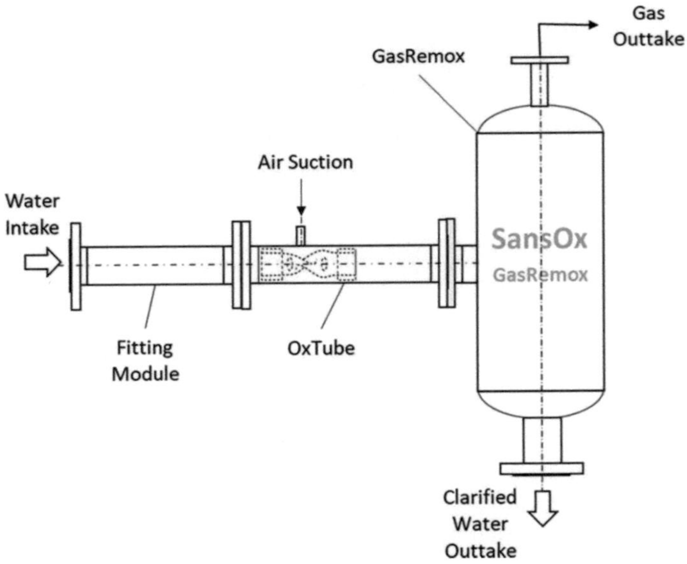

Figure 11.
Water treatment unit for most of the water integrated clarification cases. It consists of an OxTube machine, a gas removal GasRemox and a fitting module. A 60% microbe reduction was achieved with air suction, and a 100% reduction was achieved with ozone feed.

Disinfection of Hospital Wastewater/THL Summary April 3, 2018

Sample	E.coli PMY/ml	Log-red	Reduction %	MS2 PFU/ml	Log-red	Reduction %	T °C	EC µS/cm	pH	Turbity NTU	O_3 Res mg/l	O_2 mg/l
Test Water	2			1			12,8	283	7,73	0,68	0,03	10,26
Inoculum	2,500,000			870,000								
0 mgO_3/l	1,700,000		32,00	340,000		60,92	12,4	285	7,88	0,46	0,02	11,65
1,43 mgO_3/l	<0,001 (MR.)	>9,23	99,99,999,994	<2 (MR.)	>5,27	>99,9995	13,2	285	7,81	0,31	0,10	> 20
1,27 mgO_3/l	<0,001 (MR.)	>9,23	99,99,999,994	<2 (MR.)	>5,27	>99,9995	12,6	294	7,83	0,27	0,14	> 20
1,11 mgO_3/l	<0,001 (MR.)	>9,23	99.99999994	<2 (MR.)	>5,27	>99,9995	12,9	288	7,73	0,44	0,10	> 20
0,95 mgO_3/l	<0,001 (MR.)	>9,23	99,99,999,994	<2 (MR.)	>5,27	>99,9995	14,3	279	7,84	0,37	0,10	> 20
0,79 mgO_3/l	<0,001 (MR.)	>9,23	99,99,999,994	<2 (MR.)	>5,27	>99,9995	14,0	289	7,85	0,44	0,08	> 20
0,61 $mgO_3/1$	<0,001 (MR.)	>9,23	99,99,999,994	<2 (MR.)	>5,27	>99,9995	14,0	287	7,74	0,40	0,07	> 20

Disinfection of Hospital Wastewater/THL Summary April 3, 2018

Sample	E.coli PMY/ml	Log-red	Reduction %	MS2 PFU/ml	Log-red	Reduction %	T °C	EC µS/cm	pH	Turbity NTU	O_3 Res mg/l	O_2 mg/l
Test Water	2			1			12,8	283	7,73	0,68	0,03	10,26
Inoculum	2 500 000			870 000								
0 mgO_3/l	1 700 000		32,00	340 000		60,92	12,4	285	7,88	0,46	0,02	11,65
1,43 mgO_3/l	<0,001 (MR.)	>9,23	99,99999994	<2 (MR.)	>5,27	>99,9995	13,2	285	7,81	0,31	0,10	> 20
1,27 mgO_3/l	<0,001 (MR.)	>9,23	99,99999994	<2 (MR.)	>5,27	>99,9995	12,6	294	7,83	0,27	0,14	> 20
1,11 mgO_3/l	<0,001 (MR.)	>9,23	99,99999994	<2 (MR.)	>5,27	>99,9995	12,9	288	7,73	0,44	0,10	> 20
0,95 mgO_3/l	<0,001 (MR.)	>9,23	99,99999994	<2 (MR.)	>5,27	>99,9995	14,3	279	7,84	0,37	0,10	> 20
0,79 mgO_3/l	<0,001 (MR.)	>9,23	99,99999994	<2 (MR.)	>5,27	>99,9995	14,0	289	7,85	0,44	0,08	> 20
0,61 mgO_3/l	<0,001 (MR.)	>9,23	99,99999994	<2 (MR.)	>5,27	>99,9995	14,0	287	7,74	0,40	0,07	> 20

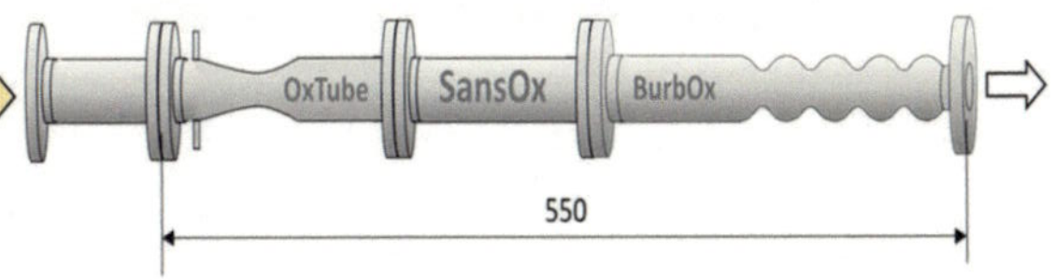

Table 1.
Disinfection efficiency of the integrated water clarification of the OxTube: 100% reduction of E. coli and MS2 was achieved with an O_3 feed of 2 l/min within 0.7 s, an increase in the oxygen concentration over the MR of 20 mgO_2/l, and no significant ozone residues/Savonia University of Applied Sciences in Kuopio [5].

DOI: http://dx.doi.org/10.5772/intechopen.1004716

Pharmaceutics and other substances	OxTube Ozone Treatment			Present Method 1		Present Method 2	
	Initial µg/l	Residue µg/l	Reduction %	Resudue µg/l	Reduction %	Residue µg/l	Reduction %
Cetirizine	5,8	0,006	99,9	4	<20	4	<20
Benzotriazole	2,8	0,896	68	1	63	2	<20
Furosemide	1,8	0,054	97	1,4	<20	1,4	<20
Hydrochloricazide	1,8	0,162	91	1,4	<20	1,4	<20
Diclofenac	1,5	0,004	99,7	1,2	<20	1,2	<20
Lamotrigine	1,4	0,756	46	1,1	<20	1,1	<20
Losartan	1,2	0,005	99,6	1	<20	1	<20
Losartan	1,2	0,246	79,5	0,7	<20	0,7	<20
5-methylbenzotriazole	0,83	0,191	77	1,4	<20	1,4	<20
Diatrizoate	0,70	0,147	79	0,6	20	0,5	<20
Atenolol	0,56	0,118	79	0,42	<20	0,35	22
Carbamazepine	0,53	0,048	91	0,4	<20	0,4	<20
Clozapine	0,53	0,053	90	0,4	<20	0,4	<20
Bisoprolol	0,50	0,105	79	0,1	77	0,4	<20
Tramadol	0,43	0,043	90	0,3	<20	0,3	<20
Citalopram	0,42	0,013	97	0,2	<20	0,2	<20
Piperacillin	0,29	0,009	97	0,32	48	0,2	<20
4-acetamidoantipyrine	0,29	0,015	95	0,15	21	0,15	<20
Azithromycin	0,20	0,012	94	0,14	<20	0,14	<20
Mirtazapine	0,17	0,006	96,4	0,08	31	0,11	<20
Atorvastatin	0,14	0,013	91	0,09	55	0,09	66
Propanolol	0,11	0,010	91	0,09	<20	0,09	<20
Quetiapine	0,11	0,006	95	0,08	<20	0,08	<20

Pharmaceutics and other substances	OxTube Ozone Treatment			Present Method 1		Present Method 2	
	Initial µg/l	Residue µg/l	Reduction %	Resudue µg/l	Reduction %	Residue µg/l	Reduction %
Naproxen	0,10	0,020	80	0,07	<20	0,07	<20
Desloratadine	0,081	0,022	73	0,1	25	0,09	33
4-formylaminoantipyrine	0,077	0,022	71	0,04	52	0,06	<20
Bezafibrate	0,069	0,014	79	0,01	78	0,05	<20
Fluvastatin	0,068	0,011	84	0,07	26	0,08	<20
Metronidazole	0,064	0,035	45	0,04	23	0,05	<20
Sotalol	0,064	0,010	84	0,06	<20	0,06	33
Sulfamethoxazole	0,063	0,011	83	0,05	<20	0,05	<20
Ketoprofen	0,059	0,042	28	0,03	62	0,02	80
Sertraline	0,05	0,011	79	0,03	<20	0,03	<20
Sulfadiazine	0,042	0,013	70	0,03	<20	0,03	<20
Clarithromycin	0,033	0,011	68	0.03	<20	0,03	24
Ramipril	0,031	0,008	74	0,02	<20	0,02	<20
Amiloride	0,023	0,014	38	0,01	<20	0,01	<20
Roxithromycin	0,008	0,003	62	0,01	<20	0,01	<20
Xylometazoline	0,013	0,002	86	0,006	<20	0,006	<20
Warfarin	0,007	0,020	<20	0,02	24	0,02	<20
Caffeine		0,018	<20	0,018	<20	0,011	52
Verapamil		0,010	<20	0,01	<20	0,01	24
Total Drug Emission Load Reduction	**24,162**	**3202**	**86,75**	**17,194**	**28,83**	**18,317**	**24,18**

Complex molecular structures of the pharmaceutics are split in separation phase before activation and clarification [4].

Table 2.
Removal of pharmaceutics and some other substances from the purified city wastewater of Kuopio Finland with OxTube integrated clarification treatment within a second.

was clarified and oxygenated at the same time as the drug residues were removed. The present wastewater purification systems separate only 1/4 of the OxTube performance. The removal of drugs by the present systems and filtering leads residues to ground or back to water, which is unacceptable. Many chemicals, such as DDT, aldrin, chlordane, chloroform, and hexachlorobenzene, have been used for decades and have led to the natural and food chain. The present treatment systems are not able to eliminate these chemicals, similar to drug residues.

The heavy gas removal facility is presented in **Figure 12**. It is applied for radon gas removal in the intake of ground water. The performance of Rn removal can be seen in **Table 3**. Two WTU facilities with a compressor were installed in the present ground water intake after the pump. The intake water was clarified and aerated at the same time as Rn removal.

5. Combined water clarification, disinfection and particle removal by flotation

The OxTube Machine can be integrated into VoxFlot flotation, as shown in **Figure 13**. The OxTube Machine generates many bubbles and even mixtures of water and bubbles under hermetic tube conditions, as illustrated in **Figure 9**. The probability of particles and bubbles being met is high, and particle attachment already occurs in the OxTube and subsequent tubes before the flotation cell [1, 3, 4, 6]. Two head vortex flows of the water and bubble mixture are generated in the VoxFlot flotation cell, which improves the particle attachment and extends the bubble path to the surface. The Bubble Trap speeds the vortex flow and pushes out a foam with particles on to the water surface. The clarified water is removed behind the bubble trap where no bubbles or particles exist.

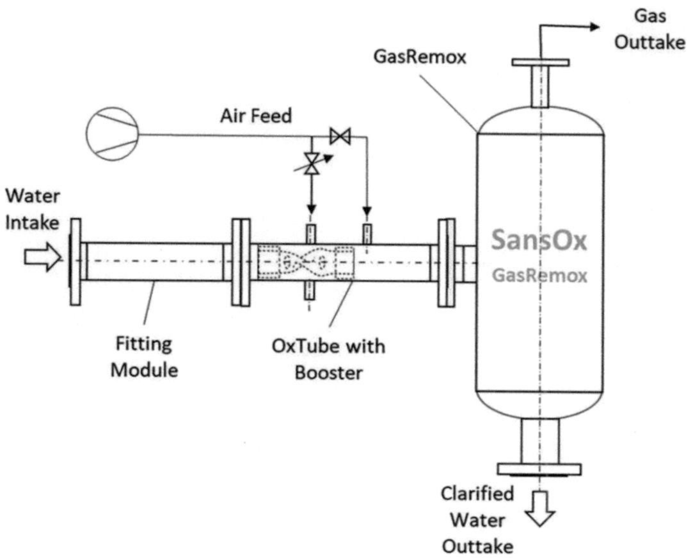

Figure 12.
Water treatment unit for heavy gas removal, such as Rn and CO_2. It consists of an OxTube machine with a booster, a gas removal GasRemox and a fitting module. Compressed air is used for separation, heavy gas blow-up and water clarification.

Water flow	Date of Sample	Monitoring Data						Rn Removal Bq/l			
		Pressure psi					Raw Water	Treated Water	Reduction Bq/l	Reduction %	Acceptance
		Pump out	OxTube 1	OxTube2	Distr. Line	Δ					
	1809–19	60	52	48	60	0	55	9	46	83.6	Passed
	1909–1919	60	52	48	60	0	50	8	42	84.0	Passed
	2309–19	60	52	48	60	0	53	4	49	92.5	Passed
	2409–19	60	52	48	60	0	52	5	47	90.4	Passed
	2509–19	60	52	48	60	0	51	4	47	92.2	Passed
	2609–19	60	52	48	60	0	47	5	42	89.4	Passed
	3009–19	60	52	48	60	0	43	8	35	81.4	Passed
62.5 m^3/b	0110–19	60	52	48	60	0	52	7	45	86.5	Passed
	0210–19	60	52	48	60	0	45	6	39	86.7	Passed
	0310–19	60	52	48	60	0	47	6	41	87.2	Passed
	0710–19	60	52	48	60	0	57	8	49	86.0	Passed
	0810–19	60	52	48	60	0	56	6	50	89.3	Passed
	0910–19	60	52	48	60	0	50	5	45	90.0	Passed
	1410–19	60	52	48	60	0	42	6	36	85.7	Passed
	1510–19	60	52	48	60	0	52	6	46	88.5	Passed
	2810–19	60	52	48	60	0	50	7	43	86.0	Passed
Requirement							52	11		79.0	

The WTU concept is presented in **Figure 9**.

Table 3.
Radon gas removal by two WTUs with compressed air and boosters from a ground water intake.

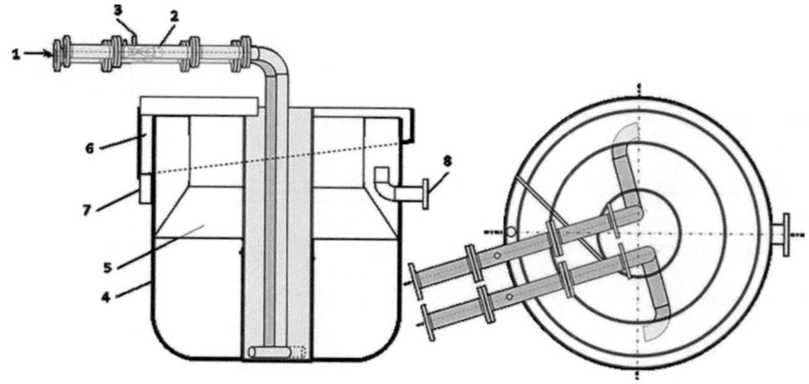

Figure 13.
VoxFlot combined clarification, disinfection and flotation; 1. Water intake, 2. OxTube water clarification, disinfection and bubble generation, 3. Gas intake, 4. Flotation cell, 5. Bubble trap, 6. Foam collector, 7. Foam outtake, 8. Clarified and disinfected water outtake. The gas fed is ozone or a mixture of air and ozone. VoxFlot is presented in Refs. [1, 3, 4].

Water clarification, disinfection, oxygenation and particle removal by flotation can be combined by feeding ozone or a mixture of ozone and air through an OxTube.

Compared to straight flow flotation with air diffusers, bubble flow through an OxTube Machine and two head vortex flow significantly decrease the unit size. Chemical feed, if necessary, is completed directly in an OxTube. Furthermore, there are no rotational components involved. These features mean reduced capital and operational costs [4].

6. Natural waters

Currently, natural waters are generally wastewater sumps. Rivers, channels, dikes and drainpipes function as drains of the natural water cycle. Ponds, lakes and seas function as sewage basins. The pollution load of waters has increased and become more hazardous, particularly in population centers where water and rain quickly drain and sewage basins during the water cycle. The energy consumption of wastewater treatment has increased significantly, but it is deficient. Evaporation is reduced tragically, which means a significant loss of climate cooling power, more floods and increased desert area. It is urgent to refresh water and food chains.

The OxTube Machine and its integrated water clarification presented in Chapter 3 can be easily utilized in many ways and positions of the water cycle. Two examples are presented in **Figures 14** and **15**, which provide an idea of how easy it could be. The Water Clarifier has all the functions of an OxTube Machine. The combined water clarifier and fountain perform integrated water clarification and landscaping by water spray. Both of these examples are easy to install and movable. Their energy consumption is found to be negligible.

Furthermore, integrated water clarification can be applied in all kinds of water transport vehicles without any additional energy consumption. Kinetic energy of the.

Vehicles in motion can be used for treatment. The flow resistance can be kept the same or even reduced by proper design. A cruising cargo ship would clarify sea water of more than 200,000 cubic meters a day at a depth of 4 meters.

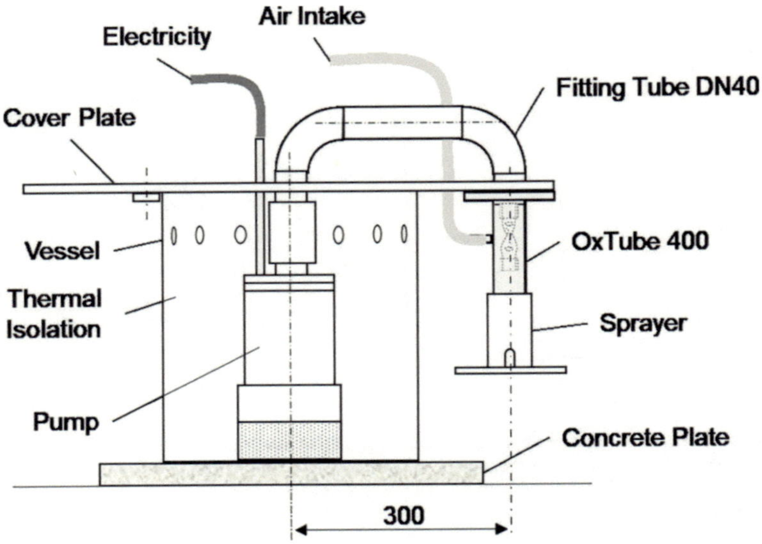

Figure 14.
The water clarifier refreshed the water with clean air without chemicals. Energy consumption is low.

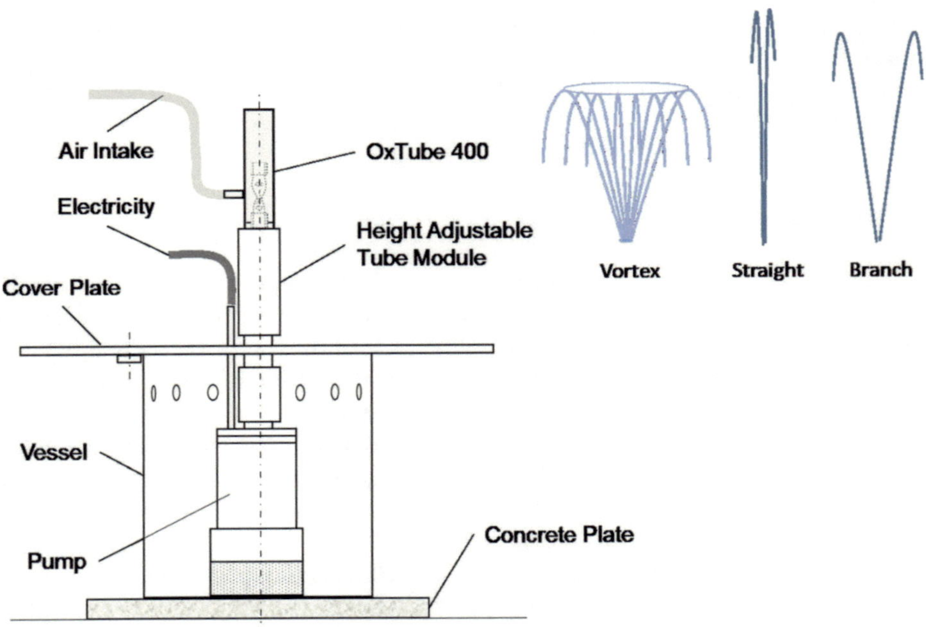

Figure 15.
Combined water clarification and fountains.

7. Conclusions

OxTube Integrated Water Clarification is found to meet the environmental and economic guiding principles and criteria listed in Chapter 1. It has been applied successfully in the removal of pharmaceuticals, radon gas, calcium, manganese and iron from various water matrices and in disinfection of hospital wastewater with ozone. More than 40 different pharmaceutical residues at concentrations greater than 24 μg/L were removed.

Moreover, 90% of the respondents used OxTube Integrated Clarification within a second in one through one run [5]. Radon removal from 57 to 8 Bq/L is achieved continuously in five ground water intake installations at water plants. Disinfection of hospital wastewater with ozone resulted in a 100% reduction in microbes and the simultaneous clarification and oxygenation of the wastewater. A 60% reduction in the amount of virus was achieved by air suction without any other gas or chemical feed. Two runs with air might result in an approximately 80% reduction. Microbe growth and variation are reduced with only clean air suction and eliminated with ozone feed.

OxTube Machine and Water Treatment Units can be installed in various positions in present treatment processes and facilities for various water matrices. They can be applied in pretreatment, removal of the water load, clarification and oxygenation and disinfection of various wastewater matrices. Furthermore, the discharge water should be clarified in terms of its environmental removal.

The four-phase clarification and bubble generation of the OxTube Machine can be applied in all the areas of water purification and recycling described in **Figure 1**. Volumes of 0.1 to 1200 m^3/h with one tube can be clarified and returned to the natural water cycle. OxTube machines can be integrated in and combined with present-day water systems. The DuOx nozzle can be integrated into hydro turbines by combining power and water treatment. Combined fountain and water treatment by the DuOx nozzle reduces energy consumption by half or more, and nature loves it according to fountain tests. DuOx Nozzle improves fire extinguishing, saves water and reduces damage caused by the addition of inert gas such as nitrogen and carbon oxide or smoke from the fire (see **Figure 4**). Watering and fertilization can be combined by means of exact spot farming. Integrated clarification should be started in water intakes because the water already contains too much waste, drug residues, radon and microplastics. Seas and oceans—refuse dumps of the water cycle—can be refreshed by integrating OxTube machines in ship structures. A cruising ship can treat 200,000 cubic meters of sea water a day without additional fuel consumption. There are more problems with water treatment than with natural water treatment.

Author details

Juhani Pylkkanen
SansOx Ltd., Toerbel, Switzerland

*Address all correspondence to: juhani.pylkkanen@t2o.fi

References

[1] Pylkkanen J. OxTube integrated water clarification for removal of pharmaceutical residues and disinfection. In: IWA Ecotechnologies for Wastewater Treatment 2021, Milan June 21-25. Conference Paper and Poester. 2021

[2] Laakso V. Test results at Oulu University [Diploma thesis]. Finland: Oulu University; 2014

[3] Pylkkanen J. New pulp and paper process water treatment system - VoxSton. China. Pulp & Paper. 2016;**35**(12)

[4] Pylkkanen J. VoxSton—A new water clarification system. In: Water Purification—Present and Future. London, UK: IntechOpen; 2023

[5] Ikonen J et al. Presence and reduction of anthropogenic substances with UV light and oxidizing disinfectants in wastewater—A case study at Kuopio, Finland. Water. 2021;**13**:360. DOI: 10.3390/w13030360

[6] Pylkkänen JH. VoxSton—New pulp and paper process water treatment system. China Pulp & Paper. 2016;**35**(12):66-70. in Chinese with English abstract

Chapter 9

Acid Mine Drainage Treatment and Control: Remediation Methodologies, Mineral Beneficiation and Water Reclamation Strategies

Sivuyisiwe Mapukata, Khuthadzo Mudzanani, Nyiko Maurice Chauke, Deogratius Maiga, Terence Phadi and Mpfunzeni Raphulu

Abstract

Although mining plays a vital role in the economic development of many countries, devastating environmental repercussions are associated with it. The extraction of mineral resources inevitably results in the generation of acid mine drainage (AMD), which entails intricate oxidation interactions that occur under ambient conditions in abandoned and active mines. The arbitrary release of AMD can lead to a series of long-term environmental problems, degradation of aquatic habitats and health complications. Over the years, extensive progress has been made in the prevention and treatment of AMD, with some processes even progressing as far as the commercialisation level. This chapter therefore discusses the process of AMD formation, preventative and control measures and AMD treatment options applicable to both operating and developed mines, as well as to researchers interested in environmental remediation and rehabilitation. Advances in mineral beneficiation and water reclamation strategies employed in the AMD treatment processes are highlighted to shed light on strides being made towards promoting a circular economy in mining industries. The featured work therefore demonstrates the global progress towards environmental protection and water resource management. The challenges and loopholes associated with the current AMD treatment methods are deliberated and possible future prospects in the field are proposed.

Keywords: acid mine drainage, mineral beneficiation, water reclamation, environmental sustainability, water treatment

1. Introduction

Although mining plays an important role in the global socio-economic development by generating wealth, providing employment opportunities and engendering foreign exchange, it often happens at the detriment of the environment. The mining industry is one of the greatest contributors to water pollution and environmental deterioration through the generation of acid mine drainage (AMD) [1]. Characterised by extremely acidic pH and high concentrations of dissolved metalloids and metals, AMD can severely contaminate surface and groundwater, as well as soils, posing a serious threat to ecological systems and human health [2]. This is because upon formation, it finds its way to the surrounding environment and water bodies through flooding of mines, runoff from open surface mining activities as well as seepage from mine residue, deposits and tailings.

Mining and mineral processing activities can generate large volumes of waste, including mill tailings, waste rock and mineral refinery waste. The exposure and oxidation of pyrites (iron disulfide) and other sulfide minerals (arsenopyrite, chalcopyrite, galena, pyrrhotite, and sphalerite) during mining and in mining waste results in the release of acidic water containing high concentrations of dissolved metals [3]. Since pyrite is the most abundant and widespread sulfide mineral, it is extensively considered as the predominant cause of AMD generation [4, 5].

The series of chemical reactions involved in the formation of AMD are shown in **Figure 1**. Briefly, when exposed to water and air, pyrite is oxidised in a reaction catalysed by microorganisms (mainly bacteria), releasing H^+, SO_4^{2-}, and Fe^{2+} ions. The Fe^{2+} ions are further oxidised to Fe^{3+} in the presence of water and air. At acidic pH, Fe^{3+} ions are hydrolysed and precipitated to form $Fe(OH)_3$, while part of the Fe^{3+} ions may continue to oxidise pyrite to sulfates and acid, forming AMD [6, 7]. The primary factors that determine the rate of acid generation therefore include; pH, oxygen concentration in the water phase; bacterial efficiency, degree of water saturation, surface area of exposed metal sulfide and the activation energy required to trigger acid generation amongst others [8].

The changes in the physicochemical properties, composition and pH of mine-impacted water deem it unsuitable for domestic, agricultural, and industrial uses [9]. Due to its distinctive bright colours, AMD alters the environmental aesthetic, causes habitat alteration and disruption of the nutrient cycle [10]. Moreover, due to its acidity, AMD induces corroding effects on infrastructure such as bridges and sculptures and contributes to the deterioration of roads around the impacted areas [11]. Additional impacts include the loss of biodiversity and decline of aquatic ecosystems [12]. Lastly, the oxidation of Fe^{2+} significantly reduces the amount of oxygen available

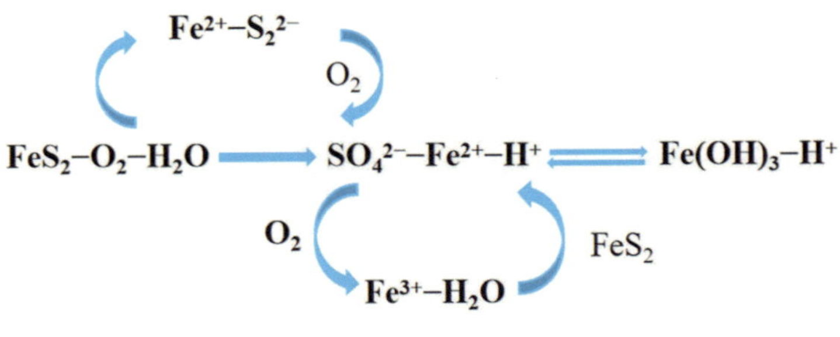

Figure 1.
Chemical reactions related to AMD formation. Reprinted with permission [6].

DOI: http://dx.doi.org/10.5772/intechopen.1003848

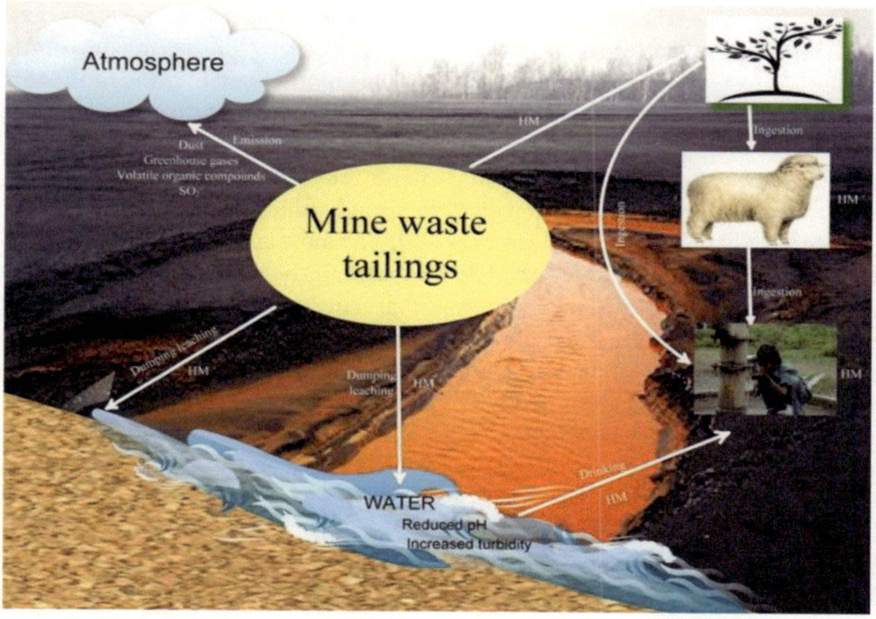

Figure 2.
Destruction of the ecosystem by mine wastes. Reprinted with permission [14].

in the water, which affects aquatic organisms [13]. The general ecological destruction caused by AMD generation is shown in **Figure 2**.

The discharge of acidic, metal and sulfate-rich water from tailings may continue for hundreds of years after mine closure if not managed properly [15]. Hence, the development of prevention and control methodologies as well as treatment technologies is vital, not only over the operational lifetime of a mine, but also during decommissioning or remediation of abandoned mine sites [16]. This will reduce the health and environmental impacts associated with AMD exposure and enhance environmental sustainability.

This chapter is thus intended to report on the current preventative, control and treatment methods applicable to operating and developed mines, as well as to researchers interested in environmental remediation and rehabilitation. The advantages and drawbacks associated with the methodologies are discussed to identify loopholes and inspire further innovative research in AMD prevention and remediation. With the aim of promoting a circular economy in mining industries, advances in mineral beneficiation and water reclamation are also deliberated. The featured work therefore highlights the global progress towards environmental protection and water resource management for enhanced ecological sustainability.

2. Prevention and control methodologies

In order to prevent the environmental damages, proper control and treatment methods need to be implemented to halt the formation of AMD. This is because, once AMD has formed and developed in mine sites, controlling its migration to the environment and surrounding water streams is rather difficult and requires costly infrastructure. For instance, Cacciuttolo et al. reported that Chile and Peru commonly use polymeric geomembranes, cut-off trenches, plastic concrete slurry walls, and grout curtain systems to control seepage at tailings storage facilities to control AMD migration [17]. Various methods have therefore been proposed to prevent the initial formation of AMD thereby retarding its environmental and health effects, mainly based on the management of rock waste and other mine tailings.

Since air, water and bacteria are key ingredients required for AMD formation, methods for controlling the formation of AMD usually function by limiting the exposure of mine areas and tailings to these resources [18]. The most commonly used method of control is minimising the penetration of air and water into the tailings waste by using dry covers; which are inorganic mineral materials that reduce the oxidation rate of sulfide minerals [19]. Dry covers are generally composed of materials with different granulation characteristics including compacted clay, portland cement and fly ash amongst others [20]. However, extensive analysis needs to be conducted during the selection of dry covers because alkaline covers like fly-ash can cause metal mobilisation from the waste, resulting in weathering of the tailing waste [21]. Alternatively, wet covers (water) have been proposed and used as an oxygen diffusion barrier. This is because, under normal conditions, the solubility of oxygen in water is low, which limits the rate of sulfide mineral oxidation when mine waste is covered with it [20].

The use of bactericides has also been reported to inhibit acid generation during hard rock and coal mining operations, resulting in an improvement in the quality of the reclaimed water [22]. Bactericides are however only effective on fresh tailings and are short-lived, and therefore do not serve as a permanent solution to AMD control [5]. Although they also have a short lifespan, organic materials such as wood waste have also been applied to cover mine tailings and the surface of mine waste, thereby preventing oxygen entry [23].

Lastly, the diversion of surface water flowing towards the mine tailings, prevention of groundwater infiltration into the tailings site and controlled placement of acid-generating waste can reduce the formation of AMD [8]. Even with such prevention measures in place, there is still a lot of mine- impacted water being generated globally due to the difficulty of maintaining some of these technologies overtime, thus research interest in the treatment of AMD has soared.

3. Treatment methods and technologies

Ideally, AMD treatment methods and technologies should have a high sulfate and metal removal efficiency, high water quality generation, low operational and labour costs, easy process design and control, minimal waste generation and low maintenance. As discussed next, several techniques have been employed for the removal of some chemical species from AMD. A description of the operation processes is provided, along with the primary advantages and disadvantages associated with the processes.

3.1 Adsorption

Adsorption is a surface phenomenon wherein atoms, ions, or molecules from gases, liquids, or dissolved solids adhere to the active sites of an adsorbent, forming a film of the adsorbate on its surface [24]. The adsorption technique offers several advantages, including affordability, high regeneration capacity, a wide pH range, the use of various raw materials, and a high metal binding capacity [25]. Although adsorption has been widely explored for treating AMD, there are significant drawbacks to consider; high selectivity and affinity of adsorption can hinder its effectiveness in decontaminating multi-charged wastewater like AMD. Rapid saturation leads to poor performance in highly concentrated solutions, further limiting its efficiency for AMD treatment. While the process has shown effectiveness in less concentrated solutions, it is primarily used as a polishing process. Moreover, the regenerates from the adsorption process are often highly mineralised and heterogeneous, making it challenging to obtain pure and high-quality minerals. Proper handling and disposal of regenerates also add to the overall cost [26]. Commonly used adsorbents for removing anionic and cationic pollutants through adsorption include activated carbon [27], zeolites [28], clay minerals [29], and ion exchange resins [30].

3.2 Chemical precipitation

Precipitation techniques are widely employed in the treatment of industrial wastewater, effluent water, and water affected by mining due to its simplicity [31]. The process entails the addition of suitable precipitating agents to the contaminated water, such as iron salts, limestone, lime, alum, or polymers. Chemical coagulants are used as a preliminary step to destabilise fine suspended particles and colloids, which then agglomerate into larger flocs. These precipitating agents react with heavy metal ions, converting them into insoluble precipitates, such as metal carbonates, metal hydroxides, and metal phosphates that can be easily separated from the water [32]. There are two main types of precipitation mechanisms used namely neutralisation precipitation and oxidative precipitation. Neutralisation precipitation involves the addition of alkaline materials to raise the pH of the water, making it more basic. On the other hand, oxidative precipitation involves the addition of oxidising agents into the contaminated water to convert the metal ions into insoluble forms [33]. The oxidising agents such as hydrogen peroxide (H_2O_2) facilitate the conversion of metal ions from their soluble reduced state to a less soluble oxidised state [34].

Both neutralisation and oxidative precipitation mechanisms can be effective in removing various heavy metals, such as lead, copper, zinc, and others, from industrial wastewater before its discharge into the environment. The choice of the specific precipitation mechanism and the type of alkaline material or oxidising agent used depends on the nature of the contaminants and the water chemistry involved in each particular case [35]. Following this step, the water is subjected to either sedimentation or filtration processes to separate the formed precipitates from the contaminated water. In sedimentation, the formed precipitates settle down under the influence of gravity, and the clear water can be decanted from the top [24, 36]. In filtration, the water is passed through a filter medium to separate the solid precipitates from the liquid phase [37]. In some cases, a secondary treatment step may be required to further purify the water. Once the water is sufficiently purified, it can be safely discharged into the environment if it meets regulatory standards. Alternatively, in water-scarce regions or in industries where water conservation is essential, the purified water can be recycled and reused for various purposes, reducing the overall water demand. This method offers several advantages for removing various metal ions from AMD. It proves to be effective in recovering valuable metals in the form of precipitates and can be optimised according to the specific pollutants present. However, there are certain drawbacks associated with this method. For instance, it leads to the generation of additional waste products. Moreover, pH adjustment may be required to achieve optimal precipitation, demanding careful monitoring. Additionally, the effectiveness of the method can be influenced by competing ions present in the water.

3.3 Filtration

Filtration is a widely used separation process that employs physical barriers to remove solid particles and contaminants from various fluids, including AMD [3, 26]. Filtration can use various types of media, depending on the specific application and the characteristics of the contaminants present in the AMD [26]. Commonly used filtration media include sand, activated carbon, diatomaceous earth, cellulose, anthracite coal, and various types of membranes such as microfiltration, ultrafiltration, nanofiltration, and reverse osmosis membranes [38, 39]. Filtration media with smaller pores can remove finer particles, while larger particles may be effectively removed by media with larger pores. Filtration processes can be categorised into two main types: depth filtration and surface filtration. Depth filtration involves the penetration of contaminants into the filter media, whereas surface filtration retains particles on the top layer of the media. Depth filtration generally offers a higher contaminant-holding capacity compared to surface filtration [40]. Depth filtration has been a critical component of water purification technology, and also in bioprocessing [40].

Prior to filtration, some AMD may require pre-treatment to optimise the filtration process. Pre-treatment steps can include pH adjustment, coagulation, flocculation, and sedimentation to facilitate the removal of specific contaminants or to prevent clogging of the filtration media [23, 41, 42]. In some filtration systems, especially those using granular media like sand or activated carbon, periodic backwashing is necessary to remove accumulated solids and prevent clogging. Backwashing involves reversing the flow of fluid through the filter to dislodge and remove trapped contaminants [43, 44]. Filtration plays a crucial role in AMD treatment and is often combined with other processes like chemical precipitation, biological treatment, and ion exchange to achieve comprehensive and effective remediation. This method offers several advantages, such as efficiently removing particulate matter and solids, which leads to improved water clarity and reduced suspended pollutants. Moreover, it offers a relatively simple implementation and maintenance process. However, it does have some drawbacks, for instance, it may not be as effective in removing dissolved pollutants. Additionally, regular maintenance and replacement of the filter media are necessary. High flow rates could also lead to reduced filtration efficiency.

3.4 Bioremediation

Bioremediation effectively treats AMD by utilising living organisms to mitigate environmental damage [45, 46]. As shown in **Figure 3**, AMD bio-remediation techniques can be classified as active and passive treatment methods [47].

Passive treatment systems use natural processes and materials to treat AMD wastewater without the need for active intervention such as mechanical or energy-intensive systems [48]. These systems typically involve constructed wetlands or ponds that promote vegetation growth and water interaction with various materials to remove contaminants. The commonly used passive treatment techniques for AMD wastewater are aerobic and anaerobic wetlands [10]. The passive treatment wetlands are designed to mimic natural wetland ecosystems and utilise plants, microbes, and soil to treat wastewater.

Unlike passive treatment, active treatment systems employ engineered processes to treat AMD. Active biological treatment of AMD involves the use of microorganisms to actively treat and remediate the acidic and metal-contaminated water.

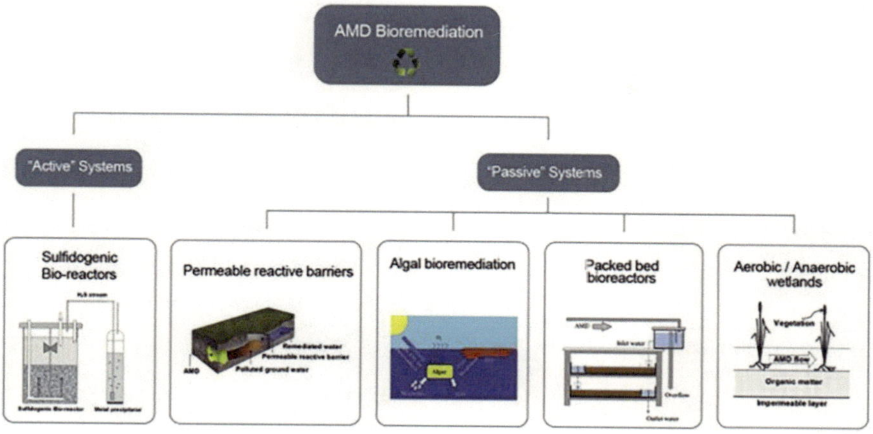

Figure 3.
Classification of AMD bio-remediation techniques: Active and passive treatment methods. Reprinted with permission [47].

This approach utilises various biological processes to neutralise acidity, remove metals, and restore water quality [47]. Generally, implementation of the active treatment for treating AMD has enhanced contaminant removal and is a more reliable and efficient treatment method than the passive [49]. On the downside, active treatment methods can be expensive to implement and maintain, they often generate a significant amount of sludge or solid waste as a by-product, which can be challenging to dispose of [26].

The efficiency of bioremediation techniques varies depending on site-specific conditions and the characteristics of the AMD [50]. A comprehensive understanding of the site and thorough monitoring are crucial for successful implementation. Additionally, bioremediation is often used in combination with other treatment methods, such as chemical precipitation or physical filtration, to achieve optimal results in treating AMD.

3.4.1 Biosoprtion

Biosorption is a bioremediation process that involves the use of living or non-living biomass to remove pollutants from contaminated water. In the case of AMD, biosorption can be employed as a method to remove toxic metals from the acidic water [50]. This process relies heavily on the ability of certain materials, such as algae, bacteria, fungi, or even agricultural waste products, to bind and accumulate metals onto their surfaces [51, 52]. These biomass materials contain functional groups like carboxyl, amino, and hydroxyl groups, which can form complex bonds with metal ions [50–52]. The metals are thus removed from the solution and immobilised onto the biomass, effectively reducing their concentration in the water. As such, different types of biomass can be used, depending on the target metals and their concentrations in the AMD [51, 52]. **Table 1** outlines some commonly used biomass materials.

The selected biomass is usually dried, ground, and sometimes chemically treated to enhance its metal-binding capacity. After the biosorption process, the metal-loaded biomass can be separated from the water through physical means, such as filtration or centrifugation [58]. The regenerated biomass may undergo a regeneration process, such as acid or alkali treatment, to restore its metal-binding capacity for future use.

Biomass Materials	Specifications	Ref.
Algae (e.g., Spirogyra, Chlorella)	• Contains functional groups (e.g., carboxyl, amino) that can bind metal ions. • Sensitive to environmental conditions (pH, temperature).	[53, 54]
Fungi (e.g., Aspergillus, Penicillium)	• High surface area biomass with numerous binding sites. • Resistant to harsh environmental conditions.	[55, 56]
Agricultural Waste Products (e.g., rice husks, coconut shells)	• Cost-effective and readily available. • Can be modified to enhance adsorption capacity. • Often used as low-cost adsorbents for various pollutants.	[57]
Activated Carbon	• High surface area with numerous micropores. • Excellent adsorption capacity for a wide range of contaminants.	[58–60]
Clay Minerals (e.g., Bentonite)	• High surface area and swelling capacity. • Suitable for metal and dye removal.	[61–63]
Natural and Modified Chitosan	• Contains amino groups for metal binding. • Used for metal and dye removal.	[64–66]
Sawdust	• Effective for dye and organic pollutant removal. • Requires modification for enhanced adsorption.	[67]
Sugarcane Bagasse	• Contains cellulose and hemicellulose for adsorption. • Suitable for metal and dye removal.	[68]
Peat Moss	• High organic matter content. • Contains humic substances for metal binding. • Used for metal and dye removal.	[69, 70]
Bark	• Natural adsorbent with porous structure and good for organic pollutant removal. • Requires pre-treatment	[69, 71, 72]
Magnetite	• Magnetic properties for easy separation. • Used for heavy metal removal.	[73, 74]
Hydrogels	• High water retention capacity. • Suitable for dye and metal removal. • Can be tailored for specific applications.	[75–77]
Polymeric Materials	• Versatile adsorbents with tunable properties. • Effective for a wide range of contaminants.	[78, 79]

Biomass Materials	Specifications	Ref.
Carbon Nanotubes	• High aspect ratio with nanoscale dimensions. • Excellent adsorption capacity for organic pollutants.	[80–83]
Soybeans Residues	• Rich in proteins and amino acids. • Requires modification for optimal adsorption.	[84, 85]

Table 1.
Frequently employed bio-based materials.

As such, biosorption is an attractive option for AMD treatment due to its cost-effectiveness and potential use of low-cost biomass materials. However, the efficiency of biosorption can be influenced by various factors, including pH, metal concentration, contact time, and biomass characteristics [52, 58, 86, 87]. Therefore, it is essential to optimise these parameters and conduct rigorous monitoring to ensure efficiency of the biosorption process in treating AMD.

3.4.2 Sulfate reducing bacteria

Sulfate-reducing bacteria (SRB) are a group of microorganisms that play a crucial role in the treatment of AMD. These bacteria can reduce sulfate ions (SO_4^{2-}) present in AMD to sulfide ions (S^{2-}), which can precipitate and immobilise heavy metals. The activity of SRB helps neutralise acidity and mitigate the environmental impact of AMD [14]. Moreover, SRB provide an effective and sustainable approach for AMD bioremediation [14]. As shown in Eq. (1), the typical SRB bioremediation process involved in the treatment of AMD involves sulfate reduction, which uses sulfate as an electron acceptor for its metabolism, converting sulfate to sulfide [88]. This metabolic activity occurs in an anoxic (oxygen-depleted) environment, typically found in the deeper layers of AMD-affected sites [14, 89].

$$SO_4^{2-} + 2(CH_2O) = \mathrm{H_2S} + 2HCO_3^{\ 2-} \tag{1}$$

The H_2S can be oxidised to sulfate or sulphur (S^o) or can react with iron sulfide minerals in the sediment as shown in Eq. (2). The insoluble metal sulfides precipitate, thus making it easier to remove them from the wastewater.

$$Fe_2\mathrm{O}_3 + 4\mathrm{S}_2\,(\text{mostly from } \mathrm{H_2S}) + 6\mathrm{H}^+ = 2FeS_2 + 3\mathrm{H_2O} + 2\mathrm{e} \tag{2}$$

The treated water undergoes settling and filtering to separate solid precipitates from the liquid phase. Monitoring and discharge ensure compliance with regulatory standards and may involve additional treatment steps, such as pH adjustment or metal removal.

The SRB can form biofilms on solid surfaces, such as rocks or sediments, in AMD affected areas [90, 91]. These biofilms provide a favourable microenvironment for the growth and activity of SRB, allowing them to establish and sustain their sulfate-reducing capabilities [91]. As such, the efficiency of SRB in AMD treatment depends on various factors, including the availability of organic matter, nutrients, and suitable environmental conditions. Optimising these factors can enhance the activity of SRB and improve their potential for treating AMD [89, 91, 92].

3.5 High density sludge (HDS) process

The high-density sludge (HDS) process has been implemented for decades as the standard in the mining industry for treating AMD, providing low metal levels in the final effluent, and reducing the generated waste sludge compared to other lime-based processes [93]. The HDS process entails addition of a neutralising reagent in the AMD to precipitate and remove metal ions from the wastewater, followed by the addition of a flocculant to aggregate fine particles, forming larger flocs. The mixture settles in a settling tank, forming a sludge layer that is concentrated with heavy metal contaminants. The treated sludge is then separated, and the clarified water can be discharged or further treated [94]. This essentially distinguishes the HDS process from other lime treatment processes i.e. the method of precipitating and separating the solids and the sludge that is formed differs. The metals can be recovered or beneficiated from the dewatered sludge. However, the current most common technique to dispose of the sludge is through landfilling, stabilisation, or recycling.

The advantage of the HDS process is that it is flexible and scalable, allowing for the treatment of various AMD sources with varying metal concentrations. The process can also be integrated into existing mine infrastructure, minimising the need for additional treatment facilities. Moreover, the process is widely accepted by regulatory agencies and has long-term stability, allowing controlled dewatering and disposal of the sludge [95].

The disadvantage of the HDS process is that it is expensive due to specialised equipment and chemicals, as well as the need for chemicals to precipitate and remove metal contaminants. It is also complex and requires skilled personnel and continuous monitoring for efficient operation [96]. Disposal of sludge, which contains concentrated heavy metals, can also be challenging and costly.

3.6 Hybrid and integrated treatment technology

Hybrid treatment technology involves combining two or more treatment methods to enhance the overall effectiveness of AMD remediation. The main objective is to take advantage of the strengths of different treatment processes to address multiple pollutants and achieve more efficient results [26]. Some common hybrid treatment approaches for AMD include a combination of chemical precipitation and filtration, biological and physical-chemical treatment, and constructed wetlands with secondary treatment [97]. Biological processes, such as bioremediation, can be combined with physical-chemical treatments to further remove contaminants from the water [98]. Constructed wetlands can be combined with secondary treatment methods like aeration or chemical addition to improve their ability to remove pollutants from AMD [49].

Integrated treatment technologies allow for a comprehensive approach to AMD treatment, targeting a wide range of contaminants [99]. One common integrated system is the multi-stage treatment system; consisting of multiple treatment stages, each targeting specific pollutants.

An example of patent technology is the Anoxic Limestone Drain/Anaerobic Bioreactor (ALD-AB) as discussed in the draft prepared by Tetra Tech Inc. [100]. The other method is membrane filtration with chemical treatment, where membrane filtration is used to remove fine particulates and residual pollutants, while chemical treatment

Technology, Combined treatment techniques	Common name
High Recovery Precipitation, Reverse Osmosis	Nafasi Water's HiPRO™
Magnesium Hydroxide and Barium Hydroxide	TUT MBO
Alkali Barium Calcium Process	CSIR-ABC
Metal Precipitation, Gypsum de-saturation, Ettringite precipitation, carbonation & recycle $AL(OH)_2$	Mintek's SAVMIN
Magnesite, Softeners, Reverse Osmosis and the Eutectic Freez	CSIR MASROE

Table 2.
Commercialised hybrid and integrated wastewater treatment technologies for AMD in South Africa.

is applied to adjust pH levels and remove dissolved contaminants. The electrochemical methods can be integrated with chemical precipitation to treat highly contaminated AMD, as they can help remove both dissolved and particulate pollutants effectively [101]. The WRC report number 940/1/04 discussed the electrochemical water treatment system called (ECODOSE) [102]. Other integrated include neutralisation combined with chemical desalination techniques i.e. the CSIR Alkaline-Barium-Calcium (ABC) and Magnesium-Barium-Hydroxide (MBO) processes. See **Table 2** for the summary of some common patented hybrid or integrated technologies in South Africa.

4. Material and mineral beneficiation strategies

The recovery and reuse of valuable resources from AMD can offset the cost of AMD remediation. Mineral beneficiation strategies therefore aim to reduce the environmental impact of AMD by treating the harmful contaminants while simultaneously recovering valuable minerals that can be reused or commercialised.

By implementing methodologies such as pH adjustments, mineral precipitation into metal hydroxides, carbonates, or sulfides can occur. This requires the use of alkaline substances which neutralise acidity by reacting with sulphuric acid, increasing the pH level [103]. The precipitated minerals or metals can then be separated from the water through sedimentation or filtration processes. Selective precipitation is another commonly used method for retrieval of valuable minerals, especially for targeted removal of specific contaminants based on their solubility characteristics [104, 105]. These mineral beneficiation processes are often used consecutively. For instance, Akinwekomi et al. recovered Fe(III) and Fe(II) for use in the synthesis of goethite, haematite, and magnetite, using sequential precipitation in batch reactors and subsequently added lime to the treated water to generate high-grade gypsum [106].

The waste generated during the treatment of AMD can also be reused to value added materials. Amanda and Moersidik generated a sludge from a precipitate which was produced from AMD active treatment ponds. They proposed that the sludge could be utilised as an adsorbent material for removing pollutants in wastewater [107]. In addition, Liu et al. effectively synthesised M-type hexaferrite from AMD sludge which could be used for absorption purposes [108].

Furthermore, in addition to the minerals, value added chemicals i.e. sulphuric acid can also be recovered form AMD as proven by Martí-Calatayud et al., who reported on the recovery of sulphuric acid from AMD by means of a 3-compartment electrodialysis [109].

5. Water reclamation

The efficiency of water treatment methods and technologies is determined by their ability to generate water with a quality level that is acceptable for reuse. Water reclamation and reuse is of great importance in the mining industry as it reduces the utilisation of the limited clean water, it also provides water that can be reused in other industries [110]. For instance, Ricci et al. implemented and reported on sequential stages encompassing microfiltration, nanofiltration and reverse osmosis to recover sulphuric acid, separate noble metals, and produce high quality reusable water from gold mining effluents [111]. Peterson et al. has also reported on the initiative driven by Battelle to optimise its HydroFlex™ water purification technology to treat AMD for the purpose of supplying non-potable water for use in hydraulic fracturing activities [112]. The reuse of water within industrial processes is therefore fast becoming a norm as laws and regulations concerning environmental protection and water security are becoming increasingly stringent.

Water reclaimed from AMD treatment can however also be used for agricultural purposes (irrigation), or even polished to generate potable water [113].

5.1 Water reclamation for irrigation

Unfortunately, due to water scarcity, many agricultural areas make use of AMD contaminated streams, with elevated trace metal concentrations, for irrigation. For instance, Garrido et al. investigated the effects of irrigation using AMD contaminated water on soil and crops (potatoes). They found that the irrigation water contaminated the soil and potatoes with high concentrations of various ecotoxic metals. The levels of Cd, Pb, and Zn in the generated potatoes exceeded commercially sold vegetable guidelines, possibly presenting a potential health risk to subsistence farmers and other consumers [114]. Many researchers have therefore resorted to treating the AMD before embarking on such endeavours.

For instance, Martins et al. worked on generating water for irrigation from AMD by comparing the efficiencies of two bioremediation systems based on sulfate-reducing bacteria. System I consisted of a packed bed reactor containing calcite tailing followed by an anaerobic packed bed bioreactor. System II consisted of a settler, fed with AMD and treated water recycle, and a sulphidogenic anaerobic packed bed bioreactor, fed with clarified settler effluent. They reported that, in addition to not needing a neutralising agent, the system II had very high sulfate reduction (>99.0%) capacity, reaching levels much lower than legislated limits for irrigation water [115].

Additionally, Kalombe et al. constructed a pilot plant for the treatment of AMD using coal fly ash. The concentration of major contaminants (sulfate, Al, Fe, Ca, Mg), and minor contaminants in the treated AMD was significantly lowered (between one and four orders of magnitude) compared to the raw AMD with at least 66.6% (728.56 kg) of treated water being recovered. Their treated water met the target water quality range (TWQR) limit for agricultural irrigation in South Africa [116].

Moreover, Han et al. recently reported on the use of fertiliser driven forward osmosis (FDFO) to extract fresh water from AMD for irrigation using novel thin film nanofibrous composite (TFNC) membranes with selective layers. They were able to generate water that promoted plant growth without posing any potential health threat [117].

5.2 Water reclamation for drinking water

The generation of potable water from AMD often requires complex treatment systems which involve filtration. For instance, Masindi generated drinking water from AMD by using magnesite to recover all the metals and increase the pH ≥ 9, lime to remove ≥ 90 of residual sulfate and soda ash to remove the residual calcium to form limestone. He further bubbled the treated water with CO_2 to remove residual sulfates and polished the treated water using a simulated Reverse Osmosis (RO) system to produce water that meets the drinking water quality as stipulated by the South African National Standard (SANS) 241 standards. The pH of the recovered drinking water was ≈ 6.5 and the metals removal efficiency of the RO system was $\approx$ 100% [118].

Additionally, Shingwenyana et al. reported on a developed and patented process by the Council for Scientific and Industrial Research (CSIR) for the treatment of AMD using an integration of a number of technologies. Their technology can reclaim drinking water while generating a number of valuable minerals such as goethite, haematite, magnetite, gypsum and limestone [119]. Lastly, Hutton et al. reported on the recovery of potable water from AMD using a Keyplan HiPRO process from the eMalahleni Water Reclamation Plant that was commissioned 2007. Although membrane health and performance are paramount in keeping this process going, they had targeted on generating 25ML/day of drinking water [120].

6. Challenges and future prospects

The closure and abandonment of mines is not usually caused by total depletion of resources, hence the continued illegal mining activities. Mine closure is typically associated with falling financial returns based on metal values, or social, political, and environmental restrictions [121]. The lack of stringent regulations for controlling abandoned mines however results in heightened environmental distress through many factors including AMD formation. Policy regulation, mine closure planning, rehabilitation and sealing of open shafts and sustainable practices are therefore essential for minimising AMD generation and preserving global water resources and ecosystems for future generations. Moreover, strengthening and enforcing environmental regulations related to mining and AMD discharge is essential to reduce pollution from operational mines and to hold responsible parties accountable.

The complexity of AMD makes its complete treatment using a single technique nearly impossible, therefore integrated treatment approaches are often implemented to generate reusable water. Essentially, addressing AMD therefore requires a multifaceted approach that accounts for the challenges associated with each technique while leveraging ongoing research and technological advancements to create more effective and sustainable treatment methods. Advancements in materials science, reactor design, and process optimisation however offer promising future prospects for more effective and sustainable AMD treatment methods. Moreover, collaborations between government entities, mining companies, environmental organisations, research associations and local communities are essential for the implementation of preventative and treatment solutions of AMD.

Ideally, mine management should consider sustainability, in which ecological restoration is an important part. Recently, Qi et al. suggested combining vegetation restoration with the covering of mine tailings as a synergistic technology for both pollution control and ecological restoration [122]. Also, with advances in AMD predictive tests and methodologies and integration with laboratory tools to measure mineralogical, geochemical, textural and geometallurgical properties, effective waste prevention and management during operation can be achieved, ultimately leading to less costly mine closure outcomes.

Although some AMD treatment processes give rise to new waste streams (e.g. brines and gypsum), finding useful applications and enhancing commercial value for products of AMD remediation is therefore an area worth investing in. For instance, Hedin et al. reported on the use of iron oxide sludge recovered from a drainage channel at an abandoned coal mine in the production of pigments, a venture which proved to be commercially lucrative [123]. Some researchers have also considered AMD as a potential low-cost raw material for the production of iron oxide particles (goethite, ferrihydrite, and magnetite), for potential application in industrial wastewater treatment [124]. Lastly, the high cost implications, energy requirements and low yield in the quantity and quality of purified water hamper the AMD treatment processes. It would thus be beneficial to invest in developing innovative technologies that will cater to these shortfalls.

7. Conclusions

This chapter has demonstrated the available technologies for the prevention, control and treatment of AMD and the urgency of furthering research and innovation in order to develop new technologies. With limitations such as high costs, low efficiency and selectivity as well as high by product (waste) generation in the currently employed methods of treatment, investments into innovative treatment processes is required. Stringent policy development and implementation is also required to regulate and limit the AMD released into the environment as well as promote a circular techno-economic strategy in the mining industry.

Acknowledgements

The authors wish to acknowledge Mintek for supporting this work as well as the Department of Science and Innovation (DSI), which supported this work through the DSI/Mintek, Nanotechnology Innovation Centre (NIC).

Conflict of interest

The authors declare no conflict of interest.

Author details

Sivuyisiwe Mapukata[1*], Khuthadzo Mudzanani[2], Nyiko Maurice Chauke[3], Deogratius Maiga[2], Terence Phadi[2] and Mpfunzeni Raphulu[3]

1 Nanotechnology Innovation Centre, Advanced Materials Division, Mintek, Johannesburg, Gauteng, South Africa

2 Measurement and Control, Mintek, Johannesburg, Gauteng, South Africa

3 Advanced Materials Division, Catalysis Group, Mintek, Johannesburg, Gauteng, South Africa

*Address all correspondence to: sivuyisiwem@mintek.co.za

References

[1] RoyChowdhury A, Sarkar D, Datta R. Removal of acidity and metals from acid mine drainage-impacted water using industrial byproducts. Environmental Management. 2019;**63**(1):148-158

[2] Luís AT, Córdoba F, Antunes C, Loayza-Muro R, Grande JA, Silva B, et al. Extremely acidic eukaryotic (micro) organisms: Life in acid mine drainage polluted environments—Mini-review. International Journal of Environmental Research and Public Health. 2021;**19**(1):376

[3] Simate GS, Ndlovu S. Acid mine drainage: Challenges and opportunities. Journal of Environmental Chemical Engineering. 2014;**2**:1785-1803

[4] Murphy R, Strongin D. Surface reactivity of pyrite and related sulfides. Surface Science Reports. 2009;**64**(1):1-45

[5] Sahoo PK, Kim K, Equeenuddin SM, Powell MA. Current approaches for mitigating acid mine drainage. Reviews of Environmental Contamination and Toxicology. 2023;**226**:1-32.

[6] Yuan J, Ding Z, Bi Y, Li J, Wen S, Bai S. Resource utilization of acid mine drainage (AMD): A review. Water (Switzerland). 2022;**14**:2385, 1-15

[7] Buzzi DC, Viegas LS, Rodrigues MAS, Bernardes AM, Tenório JAS. Water recovery from acid mine drainage by electrodialysis. Minerals Engineering. 2013;**40**:82-89

[8] Akcil A, Koldas S. Acid mine drainage (AMD): Causes, treatment and case studies. Journal of Cleaner Production. 2006;**14**:1139-1145

[9] Denicola DM, Stapleton MG. Impact of acid mine drainage on benthic communities in streams: The relative roles of substratum vs. aqueous effects [Internet]. Available from: www.elsevier.com/locate/envpol

[10] Skousen J, Zipper CE, Rose A, Ziemkiewicz PF, Nairn R, McDonald LM, et al. Review of passive systems for acid mine drainage treatment. Mine Water and the Environment. 2017;**36**(1):133-153

[11] Ekolu SO, Diop S, Azene F, Mkhize N. Disintegration of concrete construction induced by acid mine drainage attack. Journal of the South African Institution of Civil Engineering. 2016;**58**(1):34-42

[12] Ngure V, Davies T, Kinuthia G, Sitati N, Shisia S, Oyoo-Okoth E. Concentration levels of potentially harmful elements from gold mining in Lake Victoria region, Kenya: Environmental and health implications. Journal of Geochemical Exploration. 2014;**144**:511-516

[13] Cadmus P, Brinkman SF, May MK. Chronic toxicity of ferric iron for north American aquatic organisms: Derivation of a chronic water quality criterion using single species and Mesocosm data. Archives of Environmental Contamination and Toxicology. 2018;**74**(4):605-615

[14] Ayangbenro AS, Olanrewaju OS, Babalola OO. Sulfate-reducing bacteria as an effective tool for sustainable acid mine bioremediation. Frontiers in Microbiology. 2018;**9**:1-10

[15] Hesketh AH, Broadhurst JL, Harrison STL. Mitigating the generation of acid mine drainage from copper sulfide tailings impoundments in perpetuity: A case study for an integrated

management strategy. Minerals Engineering. 2010;**23**(3):225-229

[16] Deiana C, Giua L. This site is closed! The effect of decommissioning mining waste facilities on mortality in the long run. Journal of Environmental Economics and Management. 2023;**119**:102797, 1-21

[17] Cacciuttolo C, Pastor A, Valderrama P, Atencio E. Process water management and seepage control in tailings storage facilities: Engineered environmental solutions applied in Chile and Peru. Water (Switzerland). 2023;**15**:196, 1-31

[18] Kuyucak N. Acid mine drainage prevention and control options. In: International Mine Water Association Proceedings. Sevilla, Spain: Mine, Water and Environment; 1999. pp. 599-606

[19] Kleinmann RLP. At-source control of acid mine drainage. International Journal of Mine Water. 1990;**9**(1-4):85-96

[20] Si M, Chen Y, Li C, Lin Y, Huang J, Zhu F, et al. Recent advances and future prospects on the tailing covering technology for oxidation prevention of sulfide tailings. Toxics. 2023;**11**:11,1-13

[21] Lu J, Alakangas L, Wanhainen C. Metal mobilization under alkaline conditions in ash-covered tailings. Journal of Environmental Management. 2014;**139**:38-49

[22] Sobek AA, Rastogi V, Benedetti DA. Prevention of water pollution problems in mining: The bactericide technology. International Journal of Mine Water. 1990;**9**(1-4):133-148

[23] Kefeni KK, Msagati TAM, Mamba BB. Acid mine drainage: Prevention, treatment options, and resource recovery: A review. Journal of Cleaner Production. 2017;**151**:475-493

[24] Soliman NK, Moustafa AF. Industrial solid waste for heavy metals adsorption features and challenges; a review. Journal of Materials Research and Technology. 2020;**9**(5):10235-10253

[25] Tomiyama S, Igarashi T. The potential threat of mine drainage to groundwater resources. Current Opinion in Environmental Science & Health. 2022;**27**:100347

[26] Masindi V, Foteinis S, Renforth P, Ndiritu J, Maree JP, Tekere M, et al. Challenges and avenues for acid mine drainage treatment, beneficiation, and valorisation in circular economy: A review. Ecological Engineering. 2022;**183**:106740

[27] Tran HN, Tomul F, Ha NTH, Nguyen DT, Lima EC, Le GT, et al. Innovative spherical biochar for pharmaceutical removal from water: Insight into adsorption mechanism. Journal of Hazardous Materials. 2020;**394**:122255

[28] Abinandan S, Praveen K, Subashchandrabose SR, Venkateswarlu K, Megharaj M. Life cycle assessment for the environmental sustainability of the immobilized acid-adapted microalgal technology in iron removal from acid mine drainage. ACS Sustainable Chemistry & Engineering. 2020;**8**(41):15670-15677

[29] Netto E, Madeira RA, Silveira FZ, Fiori MA, Angioleto E, Pich CT, et al. Evaluation of the toxic and genotoxic potential of acid mine drainage using physicochemical parameters and bioassays. Environmental Toxicology and Pharmacology. 2013;**35**(3):511-516

[30] Zhou Y, Lu J, Zhou Y, Liu Y. Recent advances for dyes removal using novel adsorbents: A review. Environmental pollution. 2019;**252**:352-365

[31] Wang X, Gao Y, Jiang X, Zhang Q, Liu W. Analysis on the characteristics of water pollution caused by underground mining and research progress of treatment technology. Advances in Civil Engineering. 2021;**2021**:1-14

[32] Minas F, Chandravanshi BS, Leta S. Chemical precipitation method for chromium removal and its recovery from tannery wastewater in Ethiopia. Chemistry International. 2017;**3**(4):291-305

[33] Evangelou VP, Zhang YL. A review: Pyrite oxidation mechanisms and acid mine drainage prevention. Critical Reviews in Environmental Science and Technology. 1995;**25**(2):141-199

[34] Xiang Y, Wu P, Zhu N, Zhang T, Liu W, Wu J, et al. Bioleaching of copper from waste printed circuit boards by bacterial consortium enriched from acid mine drainage. Journal of Hazardous Materials. 2010;**184**(1-3):812-818

[35] Bhateria R, Singh R. A review on nanotechnological application of magnetic iron oxides for heavy metal removal. Journal of Water Process Engineering. 2019;**31**:100845

[36] Theodoro JDP, Lenz GF, Zara RF, Bergamasco R. Coagulants and natural polymers: Perspectives for the treatment of water. Plastic and Polymer Technology. 2013;**2**(3):55-62

[37] Masindi V, Akinwekomi V, Maree JP, Muedi KL. Comparison of mine water neutralisation efficiencies of different alkaline generating agents. Journal of Environmental Chemical Engineering. 2017;**5**(4):3903-3913

[38] Agboola O, Maree J, Kolesnikov A, Mbaya R, Sadiku R. Theoretical performance of nanofiltration membranes for wastewater treatment. Environmental Chemistry Letters. 2015;**13**:37-47

[39] Fu F, Wang Q. Removal of heavy metal ions from wastewaters: A review. Journal of Environmental Management. 2011;**92**(3):407-418

[40] Nejatishahidein N, Zydney AL. Depth filtration in bioprocessing—New opportunities for an old technology. Current Opinion in Chemical Engineering. 2021;**34**:100746

[41] Znad H, Awual MR, Martini S. The utilization of algae and seaweed biomass for bioremediation of heavy metal-contaminated wastewater. Molecules. 2022;**27**(4):1275

[42] Nkele K, Mpenyana-Monyatsi L, Masindi V. Challenges, advances and sustainabilities on the removal and recovery of manganese from wastewater: A review. Journal of Cleaner Production. 2022;**377**:134152, 1-17

[43] Korotta-Gamage SM, Sathasivan A. A review: Potential and challenges of biologically activated carbon to remove natural organic matter in drinking water purification process. Chemosphere. 2017;**167**:120-138

[44] Turan M. Backwashing of granular media filters and membranes for water treatment: A review. AQUA-Water Infrastructure, Ecosystems and Society. 2023;**72**(3):274-298

[45] Dhir B. Biotechnological tools for remediation of acid mine drainage (removal of metals from wastewater and leachate). Bio-Geotechnologies for Mine Site Rehabilitation. 2018;**1**:67-82

[46] Iatan EL. Gold mining industry influence on the environment and possible phytoremediation applications. Phytorestoration of Abandoned Mining and Oil Drilling Sites. 2021:373-408

[47] Rambabu K, Banat F, Pham QM, Ho SH, Ren NQ, Show PL. Biological remediation of acid mine drainage: Review of past trends and current outlook. Environmental Science and Ecotechnology. 2020;**2**:100024

[48] Ford KL. Passive treatment systems for acid mine drainage [Internet]. Available from: https://digitalcommons.unl.edu/usblmpub/19

[49] Ghimire U, Nandimandalam H, Martinez-Guerra E, Gude VG. Wetlands for wastewater treatment. Water Environment Research. 2019;**91**(10):1378-1389

[50] Anekwe IMS, Isa YM. Bioremediation of acid mine drainage – Review. Alexandria Engineering Journal. 2023;**65**:1047-1075

[51] Brazesh B, Mousavi SM, Zarei M, Ghaedi M, Bahrani S, Hashemi SA. Biosorption. Interface Science and Technology. 2021;**33**:587-628

[52] Singh S, Kumar V, Datta S, Dhanjal DS, Sharma K, Samuel J, et al. Current advancement and future prospect of biosorbents for bioremediation. Science of the Total Environment. 2020;**709**:135895, 1-24

[53] Tepanosyan G, Sahakyan L, Belyaeva O, Asmaryan S, Saghatelyan A. Continuous impact of mining activities on soil heavy metals levels and human health. Science of the Total Environment. 2018;**639**:900-909

[54] Almeida Â, Cotas J, Pereira L, Carvalho P. Potential role of spirogyra sp. and chlorella sp. in bioremediation of mine drainage: A review. Phycology. 2023;**3**(1):186-201 Available from: https://www.mdpi.com/2673-9410/3/1/12/htm

[55] Oyewole OA, Zobeashia SSLT, Oladoja EO, Raji RO, Odiniya EE, Musa AM. Biosorption of heavy metal polluted soil using bacteria and fungi isolated from soil. SN Applied Sciences [Internet]. 2019;**1**(8):1-8. Available from: https://link.springer.com/article/10.1007/s42452-019-0879-4

[56] Ali H, Khan E, Ilahi I. Environmental chemistry and ecotoxicology of hazardous heavy metals: Environmental persistence, toxicity, and bioaccumulation. Journal of Chemistry. 2019;**2019**:6730305, 1-14

[57] Karić N, Maia AS, Teodorović A, Atanasova N, Langergraber G, Crini G, et al. Bio-waste valorisation: Agricultural wastes as biosorbents for removal of (in)organic pollutants in wastewater treatment. Chemical Engineering Journal Advances. 2022;**9**:100239

[58] Wang B, Lan J, Bo C, Gong B, Ou J. Adsorption of heavy metal onto biomass-derived activated carbon: Review. RSC Advances [Internet]. 2023;**13**(7):4275. Available from: https://pmc/articles/PMC9891085/

[59] Lakshmi S, Baker S, Shivamallu C, Prasad A, Syed A, Veerapur R, et al. Biosorption of oxybenzene using biosorbent prepared by raw wastes of Zea mays and comparative study by using commercially available activated carbon. Saudi Journal of Biological Sciences. 2021;**28**(6):3469-3476

[60] Corral-Bobadilla M, Lostado-Lorza R, Somovilla-Gómez F, Escribano-García R. Effective use of activated carbon from olive stone waste in the biosorption removal of Fe(III) ions from aqueous solutions. Journal of Cleaner Production. 2021;**294**:126332

[61] El Ouardi Y, Lenoble V, Branger C, Laatikainen K, Angeletti B, Ouammou A.

Enhancing clay adsorption properties: A comparison between chemical and combined chemical/thermal treatments. Groundwater for Sustainable Development. 2021;**12**:100544

[62] Marouf R, Dali N, Boudouara N, Ouadjenia F, Zahaf F, Marouf R, et al. Study of adsorption properties of bentonite clay. In: Montmorillonite Clay [Internet]. London, UK: IntechOpen; 2021. Available from: https://www.intechopen.com/chapters/75461

[63] Sanavada K, Shah M, Gandhi D, Unnarkat A, Vaghasiya P. A systematic and comprehensive study of eco-friendly bentonite clay application in esterification and wastewater treatment. Environmental Nanotechnology, Monitoring & Management. 2023;**20**:100784

[64] Kong D, Foley SR, Wilson LD. An overview of modified chitosan adsorbents for the removal of precious metals species from aqueous media. Molecules. 2022;**27**(3):978. Available from: https://www.mdpi.com/1420-3049/27/3/978/htm

[65] Benettayeb A, Ghosh S, Usman M, Seihoub FZ, Sohoo I, Chia CH, et al. Some well-known alginate and chitosan modifications used in adsorption: A review. Water. 2022;**14**(9):1353. Available from: https://www.mdpi.com/2073-4441/14/9/1353/htm

[66] Nguyen LM, Nguyen TTH. Enhanced heavy metals biosorption using chemically modified chitosan coated microwave activated sugarcane baggage ash composite biosorbents. SN Applied Sciences [Internet]. 2019;**1**(12):1-7. Available from: https://link.springer.com/article/10.1007/s42452-019-1607-9

[67] Adegoke KA, Adesina OO, Okon-Akan OA, Adegoke OR, Olabintan AB, Ajala OA, et al. Sawdust-biomass based materials for sequestration of organic and inorganic pollutants and potential for engineering applications. Current Research in Green and Sustainable Chemistry. 2022;**5**:100274

[68] Ezeonuegbu BA, Machido DA, Whong CMZ, Japhet WS, Alexiou A, Elazab ST, et al. Agricultural waste of sugarcane bagasse as efficient adsorbent for lead and nickel removal from untreated wastewater: Biosorption, equilibrium isotherms, kinetics and desorption studies. Biotechnology Reports. 2021;**30**:e00614

[69] Denisova V, Tihomirova K, Mezule L. Extended use of sphagnum peat as a biosorbent for Zn(II): Repetitious sorption-desorption process. Agronomy Research [Internet]. 2018;**16**(4):1622-1629. DOI: 10.15159/AR.18.163

[70] Virpiranta H, Banasik M, Taskila S, Leiviskä T, Halttu M, Sotaniemi VH, et al. Isolation of efficient metal-binding bacteria from boreal peat soils and development of microbial biosorbents for improved nickel scavenging. Water. 2020;**12**:2000 [Internet]. Available from: https://www.mdpi.com/2073-4441/12/7/2000/htm

[71] Ighalo JO, Adeniyi AG. Adsorption of pollutants by plant bark derived adsorbents: An empirical review. Journal of Water Process Engineering. 2020;**35**:101228

[72] Devi AP, Mishra PM. Biosorption of dysprosium (III) using raw and surface-modified bark powder of Mangifera indica: Isotherm, kinetic and thermodynamic studies. Environmental Science and Pollution Research [Internet]. 2019;**26**(7):6545-6556. Available from: https://link.springer.com/article/10.1007/s11356-018-04098-7

[73] Carvalho Costa AWM, Guerhardt F, Ribeiro Júnior SER, Cânovas G, Vanale RM, de Freitas CD, et al. Biosorption of Cr(VI) using coconut fibers from agro-industrial waste magnetized using magnetite nanoparticles. Environmental Technology [Internet]. 2021;**42**(23):3595-3606. Available from: https://pubmed.ncbi.nlm.nih.gov/32266861/

[74] El-Dib FI, Mohamed DE, El-Shamy OAA, Mishrif MR. Study the adsorption properties of magnetite nanoparticles in the presence of different synthesized surfactants for heavy metal ions removal. Egyptian Journal of Petroleum. 2020;**29**(1):1-7

[75] Jamnongkan T, Mongkholrattanasit R, Wattanakornsiri A, Wachirawongsakorn P, Takatsuka Y, Hara T. Green adsorbents for copper (II) biosorption from waste aqueous solution based on hydrogel-beads of biomaterials. South African Journal of Chemical Engineering. 2021;**35**:14-22

[76] Baigorria E, Fraceto LF. Low-cost biosorbent hybrid hydrogels for paraquat remediation of water. Journal of Water Process Engineering. 2022;**49**:103088

[77] Nica I, Zaharia C, Suteu D. Hydrogel based on tricarboxi-cellulose and poly(vinyl alcohol) used as biosorbent for cobalt ions retention. Polymers. 2021;**13**(9):1444. Available from: https://www.mdpi.com/2073-4360/13/9/1444/htm

[78] Tofan L. Polymeric biomass derived adsorbents for Co(II) remediation, recycling and analysis. Polymers (Basel) [Internet]. 2022;**14**(9):1647, 1-24. Available from: https://pmc/articles/PMC9102464/

[79] Blaga AC, Tanasă AM, Cimpoesu R, Tataru-Farmus RE, Suteu D. Biosorbents based on biopolymers from natural sources and food waste to retain the methylene blue dye from the aqueous medium. Polymers. 2022;**14**(13):2728. Available from: https://www.mdpi.com/2073-4360/14/13/2728/htm

[80] De Luca P, Siciliano C, Macario A, Nagy JB. The role of carbon nanotube pretreatments in the adsorption of benzoic acid. Materials [Internet]. 2021;**14**(9):2118, 1-16. Available from: https://pmc/articles/PMC8122370/

[81] Rodríguez C, Briano S, Leiva E. Increased adsorption of heavy metal ions in multi-walled carbon nanotubes with improved dispersion stability. Molecules [Internet]. 2020;**25**(14):3106, 1-14. Available from: https://pmc/articles/PMC7397306/

[82] Hoang AT, Nižetić S, Cheng CK, Luque R, Thomas S, Banh TL, et al. Heavy metal removal by biomass-derived carbon nanotubes as a greener environmental remediation: A comprehensive review. Chemosphere. 2022;**287**:131959

[83] Fiyadh SS, AlSaadi MA, Jaafar WZ, AlOmar MK, Fayaed SS, Mohd NS, et al. Review on heavy metal adsorption processes by carbon nanotubes. Journal of Cleaner Production. 2019;**230**:783-793

[84] Fertu DI, Dragoi EN, Bulgariu L, Curteanu S, Gavrilescu M. Modeling the biosorption process of heavy metal ions on soybean-based low-cost biosorbents using artificial neural networks. PRO. 2022;**10**(3):603. Available from: https://www.mdpi.com/2227-9717/10/3/603/htm

[85] Bulgariu L, Bulgariu D. Functionalized soy waste biomass - a novel environmental-friendly biosorbent for the removal of heavy metals from aqueous solution. Journal of Cleaner Production. 2018;**197**:875-885

[86] Sethurajan M, Huguenot D, Jain R, Lens PNL, Horn HA, Figueiredo LHA, et al. Leaching and selective zinc recovery from acidic leachates of zinc metallurgical leach residues. Journal of Hazardous Materials. 2017;**324**:71-82

[87] Rissoni Toledo AG, Reyes Andrade JC, Palmieri MC, Bevilaqua D, Pombeiro Sponchiado SR. Innovative method for encapsulating highly pigmented biomass from aspergillus nidulans mutant for copper ions removal and recovery. PLoS One [Internet]. 2021;**16**(11):e0259315. Available from: https://pmc/articles/PMC8562857/

[88] Sinharoy A, Pakshirajan K, Lens PNL. Biological sulfate reduction using gaseous substrates to treat acid mine drainage. Current Pollution Reports. 2020;**6**(4):328-344

[89] Tran TTT, Kannoorpatti K, Padovan A, Thennadil S. Sulphate-reducing bacteria's response to extreme pH environments and the effect of their activities on microbial corrosion. Applied Sciences. 2021;**11**(5):2201. Available from: https://www.mdpi.com/2076-3417/11/5/2201/htm

[90] Schwarz A, Nancucheo I, Gaete MA, Muñoz D, Sanhueza P, Torregrosa M, et al. Evaluation of dispersed alkaline substrate and diffusive exchange system technologies for the passive treatment of copper mining acid drainage. Water. 2020;**12**(3):854. Available from: https://www.mdpi.com/2073-4441/12/3/854/htm

[91] Valdez-Nuñez LF, Ayala-Muñoz D, Sánchez-España J, Sánchez-Andrea I. Microbial communities in Peruvian acid mine drainages: Low-abundance Sulfate-reducing bacteria with high metabolic activity. Geomicrobiology Journal [Internet]. 2022;**39**(10):867-883. Available from: https://www.tandfonline.com/doi/abs/10.1080/01490451.2022.2087808

[92] Dong Y, Gao Z, Di J, Wang D, Yang Z, Wang Y, et al. Study on the effectiveness of Sulfate reducing bacteria to remove heavy metals (Fe, Mn, Cu, Cr) in acid mine drainage. Sustainability. 2023;**15**(6):5486. Available from: https://www.mdpi.com/2071-1050/15/6/5486/htm

[93] Aubé B, Lee D. The high density sludge (HDS) process and sulphate control. In: 10th International Conference on Acid Rock Drainage & IMWA Annual Conference. Santiago, Chile; 2015

[94] Murdock DJ, Fox JRW, Bensley JG. Treatment of acid mine drainage by the high density sludge process. Journal American Society of Mining and Reclamation. 1994;**1994**(1):241-249

[95] Coulton R, Bullen C, Williams C, Williams K. The formation of high density sludge from mine-water with low iron concentrations

[96] Lourenco M, Curtis C. The influence of a high-density sludge acid mine drainage (AMD) chemical treatment plant on water quality along the Blesbokspruit wetland, South Africa. Water SA. 2021;**47**:35-44

[97] Ryu S, Naidu G, Hasan Johir MA, Choi Y, Jeong S, Vigneswaran S. Acid mine drainage treatment by integrated submerged membrane distillation–sorption system. Chemosphere. 2019;**218**:955-965

[98] Chen J, Li X, Jia W, Shen S, Deng S, Ji B, et al. Promotion of bioremediation performance in constructed wetland microcosms for acid mine drainage treatment by using organic substrates and supplementing domestic wastewater

and plant litter broth. Journal of Hazardous Materials. 2021;**404**:124125

[99] Masindi V. Integrated treatment of acid mine drainage using cryptocrystalline magnesite and barium chloride. Water Practice Technology. 2017;**12**(3):727-736

[100] Ra TET, Ch TE. New World Mining District Response and Restoration Project. Bozeman, Montana: Tetra Tech, Inc.; 2011

[101] Chartrand MMG. Electrochemical remediation of acid mine drainage. Journal of Applied Electrochemistry. 2003;**33**(3/4):259-264

[102] Smit J, Van Zyl HC, South Africa. Water Research Commission. Electrochemical Treatment for the Removal of Sulphates from Acid Mine Drainage. South Africa: Water Research Commission; 2004. 102 p

[103] Pozo-Antonio S, Puente-Luna I, Lagüela-López S, Veiga-Ríos M. Techniques to correct and prevent acid mine drainage: A review. Dyna (Medellin). 2014;**81**(184):73-80

[104] Vaziri Hassas B, Shekarian Y, Rezaee M. Selective precipitation of rare earth and critical elements from acid mine drainage - part I: Kinetics and thermodynamics of staged precipitation process. Resources, Conservation and Recycling. 2023;**188**:106654

[105] de Silva RA, Secco MP, Lermen RT, IAH S, GEN H, Sampaio CH. Optimizing the selective precipitation of iron to produce yellow pigment from acid mine drainage. Minerals Engineering. 2019;**135**:111-117

[106] Akinwekomi V, Maree JP, Masindi V, Zvinowanda C, Osman MS, Foteinis S, et al. Beneficiation of acid mine drainage (AMD): A viable option for the synthesis of goethite, hematite, magnetite, and gypsum – Gearing towards a circular economy concept. Minerals Engineering. 2020;**148**:106204

[107] Amanda N, Moersidik SS. Characterization of sludge generated from acid mine drainage treatment plants. Journal of Physics Conference Series. 2019;**1351**(1):012113

[108] Liu M, Iizuka A, Shibata E. Acid mine drainage sludge as an alternative raw material for M-type hexaferrite preparation. Journal of Cleaner Production. 2019;**224**:284-291

[109] Martí-Calatayud MC, Buzzi DC, García-Gabaldón M, Ortega E, Bernardes AM, Tenório JAS, et al. Sulfuric acid recovery from acid mine drainage by means of electrodialysis. Desalination. 2014;**343**:120-127

[110] Englande AJ, Krenkel P, Shamas J. Wastewater treatment & water reclamation☆. In: Reference Module in Earth Systems and Environmental Sciences. Amsterdam, Netherlands: Elsevier; 2015. pp. 639-670

[111] Ricci BC, Ferreira CD, Aguiar AO, Amaral MCS. Integration of nanofiltration and reverse osmosis for metal separation and sulfuric acid recovery from gold mining effluent. Separation and Purification Technology. 2015;**154**:11-21

[112] Lane A, Beers T, Peterson R. Development and validation of an acid mine drainage water treatment process for source water. In: Proceedings of the 2nd Unconventional Resources Technology Conference. Tulsa, OK, USA: American Association of Petroleum Geologists; 2014

[113] RoyChowdhury A, Sarkar D, Datta R. Remediation of acid mine

drainage-impacted water. Current Pollution Reports. 2015;**1**:131-141

[114] Garrido AE, Condori J, Strosnider WH, Nairn RW. Acid mine drainage impacts on irrigation water resources, agricultural soils, and potatoes in potosí, Bolivia. In: 26th Annual Meetings of the American Society of Mining and Reclamation and 11th Billings Land Reclamation Symposium 2009. Lexington, Kentucky; 2009. pp. 480-493

[115] Martins M, Santos ES, Pires C, Barros RJ, Costa MC. Production of irrigation water from bioremediation of acid mine drainage: Comparing the performance of two representative systems. Journal of Cleaner Production. 2010;**18**(3):248-253

[116] Kalombe RM, Ojumu TV, Katambwe VN, Nzadi M, Bent D, Nieuwouldt G, et al. Treatment of acid mine drainage with coal fly ash in a jet loop reactor pilot plant. Minerals Engineering. 2020;**159**:106611, 1-9

[117] Han J, Chen S, You X, Liao Y, Huang JJ, Razaqpur AG. Acid mine drainage treatment by fertilizer drawn forward osmosis for irrigation. Resources, Conservation and Recycling. 2022;**186**:106574

[118] Masindi V. Recovery of drinking water and valuable minerals from acid mine drainage using an integration of magnesite, lime, soda ash, CO2 and reverse osmosis treatment processes. Journal of Environmental Chemical Engineering. 2017;**5**(4):3136-3142

[119] Shingwenyana R, Shabalala AN, Mbhele R, Masindi V. Techno-economic analysis of the reclamation of drinking water and valuable minerals from acid mine drainage. Minerals. 2021;**11**(12):1352, 1-17

[120] Water Institute of Southern Africa. Mine Water Division. International Mine Water Association. In: International Mine Water Conference, 19-23 October 2009, Pretoria, South Africa. South Africa: CSIR International Convention Centre; 2009

[121] Matshusa K, Leonard L. Exploring strategies for management of disasters associated with illegal gold mining in abandoned mines: A case study of Ekurhuleni metropolitan municipality. Jàmbá: Journal of Disaster Risk Studies. 2022;**14**(1):1-9

[122] Qi J, Zhang Y, Zhang J, Wu C, Chen Y, Cheng Z. Study on the restoration of ecological environments in mining area based on GIS technology. Sustainability (Switzerland). 2023;**15**(7):6128, 1-17

[123] Hedin RS. Recovery of marketable iron oxide from mine drainage in the USA. Land Contamination and Reclamation. 2003;**11**(2):93-98

[124] Flores RG, Andersen SLF, Maia LKK, José HJ, de Moreira RFPM. Recovery of iron oxides from acid mine drainage and their application as adsorbent or catalyst. Journal of Environmental Management. 2012;**111**:53-60